6198

THROUGH THE EYES OF HUBBLE

The Birth, Life, and Violent Death of Stars

Robert Naeye

KALMBACH
BOOKS

Printed in the United States of America

97 98 99 00 01 02 03 04 05 06   10 9 8 7 6 5 4 3 2 1

For more information, visit our web site at
http://www.kalmbach.com

Publisher's Cataloging-in-Publication
(Provided by Quality Books, Inc.)

Naeye, Robert.
    Through the eyes of Hubble: the birth, life,
and violent death of stars / Robert Naeye. — 1st ed.
    p.   cm.
    Includes index.
    ISBN: 0-913135-34-8

    1. Stars—Evolution. 2. Supernovae. I. Title.

QB806.N34 1997   523.88
                                QBI97-681

Book and cover design: Kristi Ludwig

FOREWORD

Imagine a sky without stars; a dark night with nothing punctuating it except for the Moon and planets roaming against a black, utterly featureless backdrop. How very much we would miss! There would be no fundamental references to tell us where we are, no stars to help us mark the compass points and guide us across Earth.

But that is only the beginning of the gifts presented by the stars. Step outside and admire the constellations: mighty Orion of northern winter, majestic Leo of spring, the great winged birds Cygnus and Aquila of summer, heroic Perseus of autumn. Though apparently only randomly placed dots on the sky, the constellations give us a personal way of piercing the heavy curtain of time to enter the hearts and minds of our ancient forebears, of learning about their myths, stories, dreams, and sorrows.

Perhaps more important yet is the inspiration given the aesthetic human mind by the sheer beauty of a nightly pageant that has fostered a vast amount of poetry, music, art, and philosophy. Everywhere we see sparkling gems, the brighter ones flashing color, the vast number of fainter ones disappearing into the blackness, together blending seamlessly into the Milky Way, the magnificent broad pathway made of millions of faint stars that seems to embrace Earth. Even without knowing what the stars really are, we get a sense of majesty from the grand display wheeling slowly overhead that makes us look outside ourselves to wonder about our place within the cosmos.

Equally significant is what the real physical natures of the stars tell us. Since the invention of the telescope nearly 400 years ago, astronomers have slowly pieced together a picture of the universe that shows us where we are, how we arrived here, how our Sun, Earth, and planetary neighbors were born, and where they are all headed. Modern astronomy allows us to see ourselves in the stars, as we can watch systems like ours being born and can see stars expire. Some die quietly, and others go with awesome violence, their death spilling matter and chemical elements into the interstellar void as nourishment for future generations of stars, planets, and maybe life. In the past century, we have learned that our own existence is profoundly intertwined with the stars in ways the ancients never could have imagined.

Since the early 1600s, when Galileo turned his tiny telescope to the heavens, astronomers have been building ever bigger, ever better telescopes with which to examine the stars. Earth-based telescopes, however, have a severe problem. Our atmosphere terribly degrades stellar images, giving us a frustratingly fuzzy view in even the best of our terrestrial instruments. Above almost all Earth's atmosphere flies the Hubble Space Telescope, orbiting in a near-vacuum where the stars appear as steady points. This extraordinary telescope can therefore make full use of its optical power to allow us to see fine detail and to probe into the once-private hearts of the stars.

Over its years of service, Hubble has compiled an impressive list of images that yield a vastly clearer view of the way in which stars—and perhaps even planetary systems—are born, live out their lives, and die. At the same time, the stunning colorful views return us to the aesthetics of the heavens, perhaps to inspire the poets and philosophers among us to probe yet more deeply into the nature and meaning of the sky above. Now, once again, try to imagine the sky without its stars.

James B. Kaler, University of Illinois, Urbana, Illinois

ACKNOWLEDGMENTS

There is one name on the cover, but this book is the result of the work of many people. My thanks to Michael Carroll for his invaluable contributions to this book. Every painting you see in this book is an original work of art never before published. I'd like to thank James Kaler for writing the Foreword. Thanks also to the following scientists and astronomers for their technical guidance: Adam Burrows, David Des Marais, Ronald Gilliland, David Golimowski, Ed Guinan, Alice Harding, Steve Hawley, Jeff Hester, Mario Livio, Curt Michel, Jon Morse, Roger Romani, Ed Sion, David Spergel, and David Stevenson; to illustrators Sue Biebuyck and Elisabeth Rowan; to Cheryl Gundy of the Space Telescope Science Institute's Office of Public Affairs, and Forrest Hamilton and his HST Nuggets web site.

I also want to thank my colleagues: on the *Astronomy* editorial staff—Bonnie Gordon, Dave Eicher, Rich Talcott, John Shibley, Patti Kurtz, Jeanette Brown, Terry Conley, and Julie Sherwin—who put up with my sometimes irritable mood while I was furiously working on this project; and in Kalmbach's Books Department—acquisitions editor Terry Spohn, managing editor Sybil Sosin, copy editor Mary Algozin, and designer Kristi Ludwig.

Finally, and most important, I want to thank my parents, Richard and Patricia, for their support. Without their help, the book you see would not have been possible.

CONTENTS

GREAT BALLS OF FIRE

Five billion years ago, long before the Earth and Sun were born, a star exploded. This wasn't just any star, it was a supergiant star hundreds of times larger than the Sun. For a few days, the flash of the explosion nearly equaled the combined light of all the Galaxy's 200 billion stars. The cataclysm spewed gaseous debris in all directions at speeds thousands of times faster than a bullet. A shock wave of energy ripped through the surrounding space, plowing up gas and dust as a snowplow piles up snow. The blast and radiation wreaked devastation, destroying or sterilizing any hapless planets that lay in their path.

But out of this appalling destruction were planted the seeds of new life. The gradually weakening shock wave eventually rammed into a tenuous cloud of cold gas, causing it to start contracting under the force of gravity. Tens of millions of years later, this cloud formed a newborn star. And 4.6 billion years after that, on one of the small chunks of debris left over from the formation of the star, intelligent beings emerged who could ask questions about their origin. If this scenario is correct, our planet, our Sun, our solar system, everyone we know, everyone we love, everything we ever cared about—all of these things owe their existence to a stellar cataclysm that occurred some 5 billion years ago.

Our lives are irrevocably linked to the stars. Nuclear reactions deep inside the Sun provide the sustenance for life on Earth. The oxygen we breathe, the calcium in our bones, and the iron in our blood were forged inside stars that have long since passed away. To know the stars is to know ourselves.

A Chinese proverb says that a long journey begins with a single step. In science, the road to understanding starts with a question. The modern understanding of stars can thus be traced back thousands of years, to when our ancestors looked at the night sky and wondered, "What are those points of light?"

Stars have begrudgingly yielded their secrets as the mysterious aura that once surrounded their nature has been slowly peeled away under the onslaught of modern science. Slowly, painstakingly, science has led humanity toward understanding as it chips away at the edge of the unknown. Today, we know what the stars are. Stars are gigantic spheres of hot gas—great balls of nuclear fire. We also know they are very far away, farther away than the ancients ever could have imagined. Scientists have moved beyond the most fundamental questions. As we are about to enter a new millennium, astronomers have a basic understanding of how stars are born from giant clouds of gas, how stars live by transforming light elements into heavier ones, and how they die either in relative quiescence or in spectacular explosions. But as scientists answer one set of questions, deeper ones inevitably emerge. Scientists have taken great strides in their quest to understand the stars, but they remain humbled by what's left to be learned.

Progress in astronomy goes hand in hand with advances in technology. The history of astronomy shows that improved instruments open new windows to the universe, revealing entire classes of objects never before seen. Throughout the 20th century, astronomers have built bigger and bigger telescopes, equipped with better and better detectors, allowing them to see deeper and deeper into the universe. But ground-based telescopes always suffer from a fundamental limitation: They must look through Earth's atmosphere. It's like looking through a foggy pair of eyeglasses—the atmosphere distorts and blurs light coming from celestial objects in deep space. Light traverses trillions upon trillions of miles relatively undisturbed, only to be blurred in the final 50 miles (80 kilometers) of its voyage.

From its perch 380 miles (612 kilometers) above Earth, the Hubble Space Telescope is ideally suited to unlocking many of the remaining secrets of the stars. Without an atmosphere to contend with, Hubble has given astronomers an unprecedented sharp look at star birth and star death. Where a ground-based telescope sees an amorphous smudge, Hubble sees a glorious, colorful, and complex nebula surrounding a dying star. Where a ground-based telescope sees a bright blob, Hubble sees thousands of individual stars crammed into a small region of space. As Arizona State University astronomer Paul

Scowen says, "With Hubble, we get such a clearer view of the universe that every time you get a picture, it is humanity's best view of that object, period."

The telescope has shown astronomers that star birth is a much more dynamic and violent process than they had ever imagined. It has given astronomers their sharpest view of incredibly narrow hypersonic jets of hot gas shooting away from newborn stars and stretching across trillions of miles of space. It has shown how young, massive stars engage in matricide, destroying the very gas clouds from which they were born.

Hubble has also offered new insights into stellar death. Some of the Space Telescope's most spectacular images show violent shock waves ripping through space, and multicolored layers of gas ejected by stars thousands of years before they exploded. Just as paleontologists study long-extinct creatures from their fossilized bones, astronomers can piece together the final stages of a star's life by observing its remains.

Science often advances by the development and testing of models. Astronomers develop models to explain the various phenomena they observe in space. A good model will make predictions of what should be seen when newer and better instruments become available. With ground-based telescopes, there were limits to how far many astronomical models could be tested. In some instances, a host of different models were developed to explain a particular phenomenon. Which models were right? Which ones were wrong? Hubble's discriminating eye has often allowed it to play the role of ultimate arbiter. In some cases, it spectacularly confirmed cherished theories of how the universe behaves; in other instances, it sent theorists scrambling back to the drawing board.

"This is the way science works," says astronomer John Trauger of NASA's Jet

EDWIN P. HUBBLE. The Space Telescope is named after Edwin P. Hubble (1889–1953), the greatest observational astronomer of the 20th century. In 1923 Hubble became the first astronomer to prove that there were other galaxies in the universe besides our own. In 1929 Hubble's observations led to the realization that we live in an *expanding universe*. Both discoveries revolutionized humanity's perception of the universe and our place within it.

Propulsion Laboratory. "A lot of what you thought is turning out to be true, but there's further information that you hadn't considered before. This is something that keeps happening in astronomy. Each time you try a new technique or use a larger telescope, you know for sure that you'll find something new."

Astronomers are quite understandably eager to use this unique instrument, and they compete with one another for its precious time. For every observing proposal that is accepted by the Space Telescope Science Institute, in Baltimore, Maryland, the agency contracted by NASA to operate Hubble, at least three are rejected.

While the public is primarily interested in the sheer beauty of Hubble images, Arizona State University astronomer Jeff Hester sounds the common refrain of the astronomical community: "The purpose of using the Hubble Space Telescope is not just to make pretty pictures. The purpose is to do good science." Books and magazine articles often leave unstated the fact that most of Hubble's images do not accurately portray what the human eye would see if one could travel in a spaceship and view these celestial objects up close. For example, Hubble often shoots these objects through a series of filters that allow only an extremely narrow range of colors to pass through. Each of these colors corresponds to a particular type of atom or a particular gas temperature, and the decision to assign a particular color to a particular atom is often arbitrary. Astronomers later combine these images on a computer. With their highly trained eyes, they can interpret what they see to piece together an understanding of what they are observing, which adds one more bit of knowledge to human understanding of the universe.

So yes, Hubble is satiating the public's appetite for pretty pictures. But more important, it is adding to the collective sum knowledge of humanity's understanding of our origins, our place in the universe, and the profound connection between the stars and ourselves.

STScI

BUS-SIZED 'SCOPE. The Hubble Space Telescope under construction at Lockheed Missiles and Space Company. The telescope weighs 12 tons and has a primary mirror about eight feet (2.4 meters) in diameter. Hubble's primary mirror is modest when compared to the largest ground-based telescopes, which are four times larger. But unlike telescopes on the ground, Hubble doesn't have to look through a murky atmosphere.

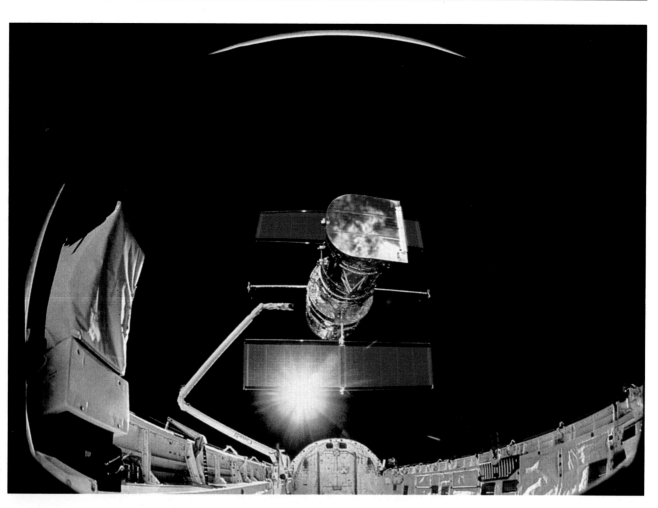

DELIVERING THE GOODS. On April 25, 1990, a day after liftoff, the space shuttle *Discovery*'s mechanical arm deployed the Space Telescope into Earth orbit.

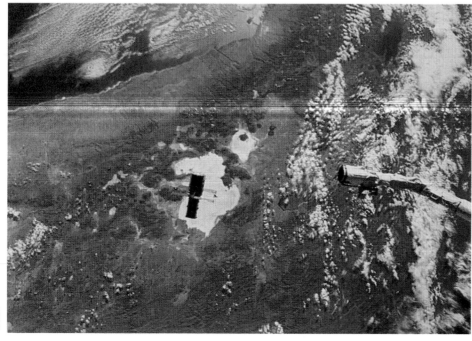

ROUND AND ROUND. Hubble orbits 380 miles (612 kilometers) above sea level. With an orbital speed of 17,500 miles (28,000 kilometers) per hour, it circles Earth every 90 minutes. Here we see the Space Telescope shortly after its deployment; the shuttle's mechanical arm is in the foreground.

11

HOUSE CALL. Hubble's primary mirror, ground to the wrong formula, was off by only 1/50 the width of a human hair. But that was enough to essentially ruin most of the telescope's primary scientific objectives. In December 1993 Space Shuttle astronauts Tom Akers, Jeffrey Hoffman, Story Musgrave, and Kathryn Thornton installed corrective optics and performed other repairs and upgrades. Here we see Story Musgrave at work.

STScI

BIG DIFFERENCE. Hubble's new optics give it a remarkably sharp view of celestial objects. In early 1994, shortly after the first servicing mission, Hubble snapped this image of the central regions of galaxy M100 in the constellation Virgo. This galaxy lies 56 million light-years away, meaning that the light we're seeing from this galaxy started its journey to Earth 56 million years ago, shortly after the age of the dinosaurs. In February 1997 shuttle astronauts installed two new sophisticated instruments.

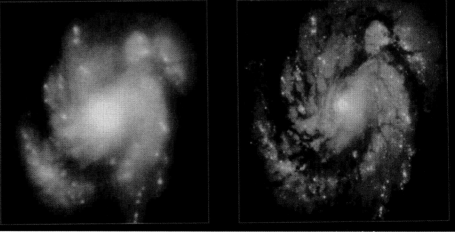

Galaxy M100 Core Comparison
Hubble Space Telescope

STScI

Tony and Daphne Hallas

CELESTIAL ILLUSION.

It is natural to assume that the brightest stars are the closest. But stars are not like lightbulbs with equal wattage. Only 9 of the 25 stars within 13 light-years of Earth can be seen with the naked eye, while several extremely luminous stars can be seen across hundreds of light-years of space. The bright blue star Rigel, at the lower right of the constellation Orion, is 50,000 times more luminous than the Sun, which makes it one of the brightest stars in the night sky, even though it is about 1,000 light-years away. The bright red star Betelgeuse at the upper left is about 500 light-years away. The Sun could only be seen with the naked eye to a distance of 60 light-years. Just as a flame changes color as it gets hotter, a star's color indicates its surface temperature. Red stars like Betelgeuse are the coolest. As Mother Nature cranks up the temperature, stars go from orange to yellow to white to blue.

THE STAR NEXT DOOR

There's one star the Hubble Space Telescope cannot observe: the one that is closest to us and dearest to our hearts. If astronomers at the Space Telescope Science Institute pointed Hubble at the Sun, the tremendous glare would instantly fry the telescope's detectors. But with the next closest star 25,000,000,000,000 miles (40 trillion kilometers) away, the Sun gives astronomers their sole opportunity to view a star up close. To understand the stars, we must first understand the Sun. Fortunately, there are plenty of superb ground-based and space-based instruments for observing the Sun. So let's forget Hubble for a few moments and turn our attention to the star that's next door—the star that gives us life.

By stellar standards, the Sun is merely average. If you think of stars as vehicles on the road, the Sun would be a station wagon: it's bigger than most cars, but the largest vehicles, the 18-wheelers, are significantly bigger!

By Earthly standards, there's nothing average about the Sun. You'd have to string 109 Earths together to span the Sun's 865,000-mile diameter (1.4 million kilometers). Driving 16 hours a day at 60 miles (96 kilometers) per hour, with no days off, it would take you $7^{3}/_{4}$ years to circumnavigate the Sun.

(Of course you'd be vaporized a nanosecond into your journey.) More than 1 million Earths could comfortably fit inside. The Sun has the mass of 333,000 Earths. Even mighty Jupiter pales in comparison, weighing in with a mass only $1/_{1,000}$ that of the Sun. The Sun, like the greedy medieval kings who lived in ornate palaces while taxing their subjects into abject poverty, contains 99.9 percent of the matter in the solar system.

But if you're searching for solar superlatives, look no further than the core—a 125,000-mile-wide inferno of unimaginable fury (200,000 kilometers—about 16 times the diameter of Earth). The sheer weight of overlying matter squeezes the core to a density of six tons (the weight of an adult male elephant) per cubic foot, more than 10 times the density of lead. The overlying material also acts as a giant pressure cooker, heating the core to a hellacious 15 million kelvins (27 million degrees F). In this seething cauldron, electrons are stripped from their host atoms. Electrons and atomic nuclei zip around at speeds of several hundred miles per second. At those velocities, particles smash into one another with tremendous force. And it's these high-energy collisions that cause the Sun—and all stars—to shine.

The key to understanding how stars shine lies in Albert Einstein's famous equation $E = mc^2$. This equation basically says that matter can be converted into energy, and vice versa. E stands for energy, m for mass, and c for the speed of light: 186,000 miles per second (300,000 kilometers per second), meaning that a photon of light could circle the Earth seven times per second. In other words, c is a *very* large number. And when you multiply a large number by itself, you wind up with a very, very, very large number. In plain English, this equation tells us that a tiny amount of mass can unleash a stupendous amount of energy, as is demonstrated every time an atomic bomb is detonated.

Here's how it works. When the Sun was born, its core was about 90 percent hydrogen, the simplest and lightest element because its nucleus consists of a single particle, a proton. The remaining 10 percent is almost all helium, the second lightest element. A single hydrogen nucleus (a proton) zipping around the Sun's core collides with another hydrogen nucleus. This initiates a series of reactions that ultimately fuses four hydrogen nuclei into one helium nucleus, which consists of two protons and two neutrons. This reaction would be like smashing four golf balls together to form a baseball. Each reaction produces energy in the form of gamma-ray photons, the most energetic form of light. This hydrogen-to-helium fusion reaction is similar to the one that gives hydrogen bombs their devastating power.

Each second, the Sun's core fuses 600 million tons of hydrogen into 595 million tons of helium. The missing 5 million tons is converted into the energy equivalent of 1 billion one-megaton hydrogen bombs. The energy produced in the Sun's core each second could fuel the American economy for 7 million years. At a distance of 93 million miles (150 million kilometers), Earth only captures one-half of one-billionth of the Sun's energy—but that's still plenty to provide nourishment for a thriving biosphere.

The fusion of hydrogen into helium is responsible for the Sun's energy output. All of this phenomenal energy is produced in the core; only in the core are the temperatures and pressures high enough to overcome the tendency of positively charged protons to repel one another. The heat emanating from the core exerts an outward gas pressure everywhere inside the Sun. This gas pressure counterbalances the inward pull of gravity, preventing the Sun from collapsing in upon itself.

The Sun is slowly fusing its hydrogen reserves into helium. At present, the Sun's core has converted roughly half of its hydrogen into helium. The Sun is almost 5 billion years old, so 5 billion years from now all of the hydrogen in the Sun's core will be gone. This will have grave consequences for the well-being of the Sun and Earth, as we shall see in Chapter 4.

A photon's journey from the core to the surface is a long and tortuous road, for like a salmon swimming upstream, it must fight its way through hundreds of thousands of miles of dense gas. Both in the core and in the region directly above it, the 185,000-mile-thick radiative zone (300,000 kilometers), photons constantly bump into free electrons. Electrons absorb photons and release many new ones at lower energy levels in various directions. Even though photons are constantly being created and destroyed, they appear to follow a zigzag path that astronomers call a "random walk." The random walk resembles the path of a drunk who stumbles out of a bar and constantly bumps into cars, trees, and telephone poles. It takes a million years—even traveling at the speed of light—for the descendants of the original gamma-ray photon to reach the outer edge of the radiative zone.

As photons make their way toward the surface, temperatures and pressures decrease. By the time they are 130,000 miles (209,000 kilometers) from the surface, temperatures are low enough that electrons attach themselves to atomic nuclei. The resulting atoms absorb the energy contained in the photons and carry it to the surface by a process known as convection.

Every time you watch a pot of boiling water, you are witnessing convection at work. Convection occurs wherever a layer of cool material overlies a layer of much hotter material. Inside the Sun an immense circulation pattern develops. Large columns of hot gas rise to the surface at speeds exceeding 200 miles (320 kilometers) per hour, while cooler material sinks to the bottom of the convection zone. These convection currents, along with the Sun's rotation, generate solar magnetic activity such as sunspots, flares, and prominences.

Once photons reach the surface, a 200-mile-wide region (320 kilometers) called the photosphere (the "sphere of light" in Greek), they're free to escape into space. By this time, the original gamma-ray photon has been transformed into thousands of visible and infrared light photons, which is fortunate for us because gamma-ray photons pack so much energy that they can rip living cells to shreds. Once photons reach the surface after their million-year journey, it takes a mere eight minutes to cross the 93-million-mile void (150 million kilometers) between the Sun and Earth.

The photosphere's 5,800-kelvin temperature (10,000 degrees F) gives the Sun its yellowish color. Although 5,800 kelvins seems like a seething cauldron by terrestrial standards, it's downright tepid when compared to the 15-million-kelvin core.

Surprisingly, the hottest parts of the Sun are situated in its wispy outer atmosphere, the corona. Recent measurements by the joint NASA/European Space Agency SOHO probe have measured coronal temperatures as high as 100 million kelvins, although most of the corona is a more temperate 2 million kelvins. But the gas is so tenuous (1.2 cubic miles of coronal gas would weigh only one pound) that you could stick your hand in a container of coronal gas without feeling a thing.

In a sense, Earth itself lies within the corona. The Sun gives off a gusty breeze of protons and electrons called the solar wind. Spacecraft have measured wind speeds topping 2 million miles per hour. The solar wind, which emanates from the corona, continues out far beyond Pluto. The solar wind causes the gas tail of comets to point away from the Sun.

Astronomers can't see directly below the solar surface, so they feed the laws of physics into sophisticated computer models, which spit out predictions that almost perfectly match observations. Astronomers are also helped by the fact that the Sun rings like a bell. They have set up elaborate networks to observe the solar oscillations round the clock, a science called helioseismology. By watching how the Sun oscillates at different frequencies, astronomers can deduce its internal structure. All of these observations and computer models are self-consistent, giving astronomers a great deal of confidence that their solar model is essentially correct. "This basic picture of the Sun has existed for decades and forms a cornerstone for understanding other stars," says solar astronomer David Hathaway of NASA's Marshall Space Flight Center.

So what is a star? In essence, a star is a gigantic nuclear fusion reactor surrounded by a thick, insulating blanket that muffles the force of the billions of H-bombs' worth of energy going off each second. This begs the question of how such a monstrous entity comes into existence. Here's a question that Hubble can help answer.

Convective zone

Radiative zone
9 million °F
5 million kelvins

Random walk
radiation pattern

Core
27 million °F
15 million kelvins

Chromosphere

Photosphere
10,000 °F
5,800 kelvins

Prominence

INTERNAL STRUCTURE OF THE SUN

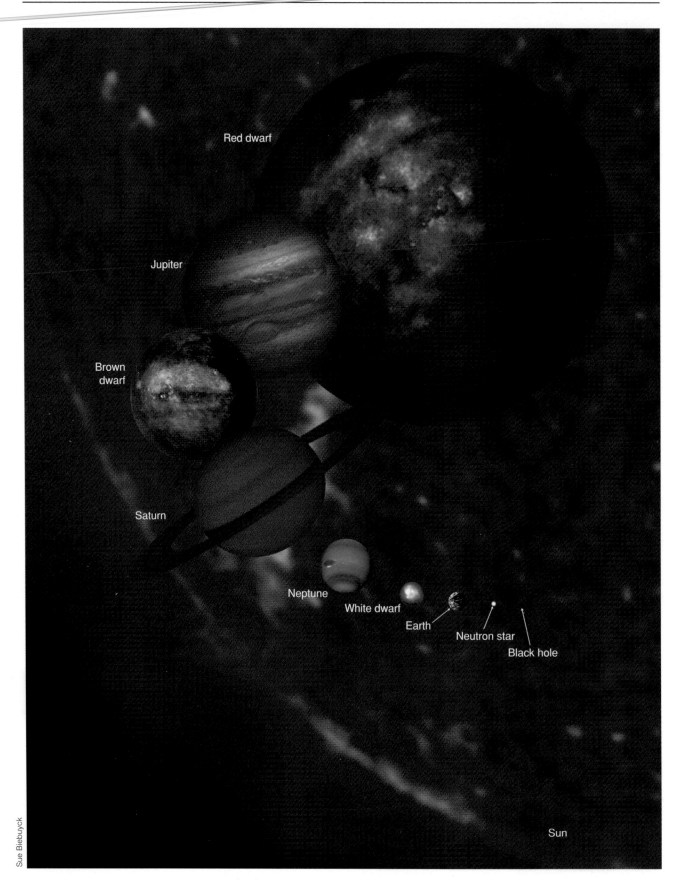

Red dwarf

Jupiter

Brown
dwarf

Saturn

Neptune

White dwarf

Earth

Neutron star

Black hole

Sun

OBJECTS SMALLER THAN THE SUN

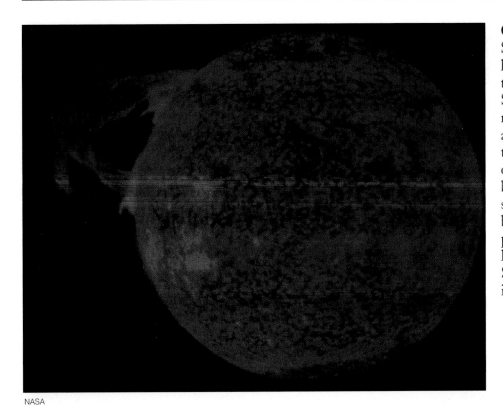

GIVER OF LIFE. The Sun is a sphere of seething hot gas. While gigantic by terrestrial standards, the Sun's size, mass, luminosity, and temperature are downright middle-of-the-road compared to other stars. The Sun is 4.6 billion years old, and should last for another 5 billion years. A giant prominence lies at upper left in this picture, which Skylab astronauts took in 1973.

PROMINENCE. Instabilities in the solar magnetic field can lift gas off the solar surface to form graceful prominences, which can hover over the Sun for days or weeks. The largest prominences easily dwarf Earth, extending for hundreds of thousands of miles.

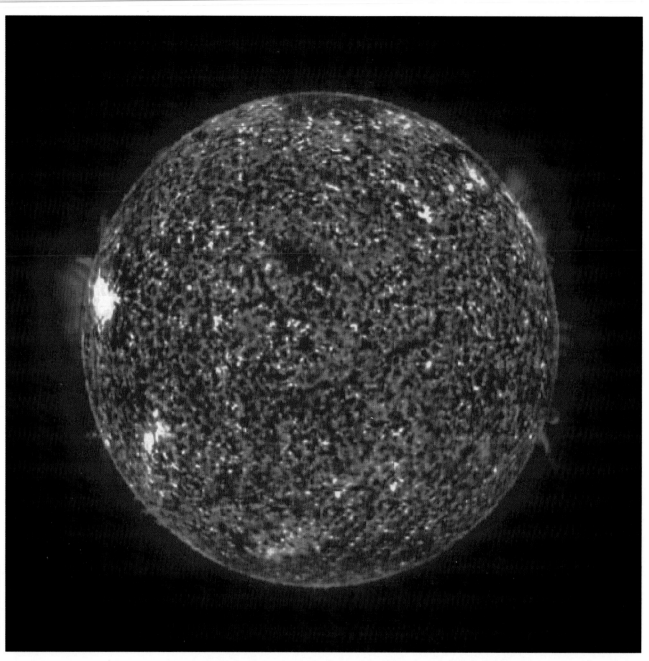

SOHO'S SUN. In December 1995 NASA and the European Space Agency launched a spacecraft called the Solar and Heliospheric Observatory, or SOHO for short. SOHO's 12 scientific instruments monitor the Sun, the corona, and the solar wind round the clock. In just the first few years of operation SOHO has returned a wealth of data, enhancing astronomers' understanding of our parent star.

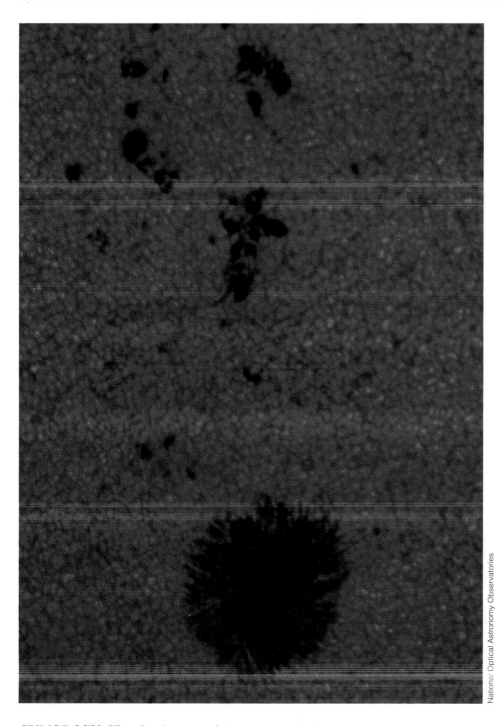

National Optical Astronomy Observatories

SUNSPOTS. The Sun's powerful magnetic field generates sunspots, which can be several times larger than Earth. Sunspots are actually extremely bright, but they appear dark because they are slightly cooler than the surrounding gas. Individual spots generally last one or two months. The number of sunspots waxes and wanes over a period averaging 11 years. The granulation pattern visible in this photo arises from circulation patterns deep inside the Sun. The bright patches represent places where hot gas is upwelling from the interior, while the black boundaries are places where cooler gas is sinking. More than 1 million granules dot the solar surface. The average granule spans 1,300 miles (2,000 kilometers), roughly half the distance across the continental United States. The granules appear and disappear in a period of 8 to 20 minutes.

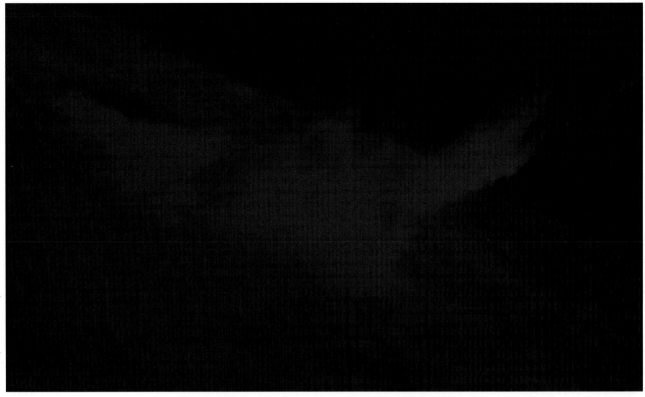

National Optical Astronomy Observatories

FLARE. The Sun is prone to sudden and violent outbursts called flares. While flares are much smaller than prominences and last only minutes to hours, they can release considerably more energy.

Randall Wehler

NORTHERN LIGHTS. A wind of charged particles emanating from the Sun slams into Earth's magnetic field, generating an aurora.

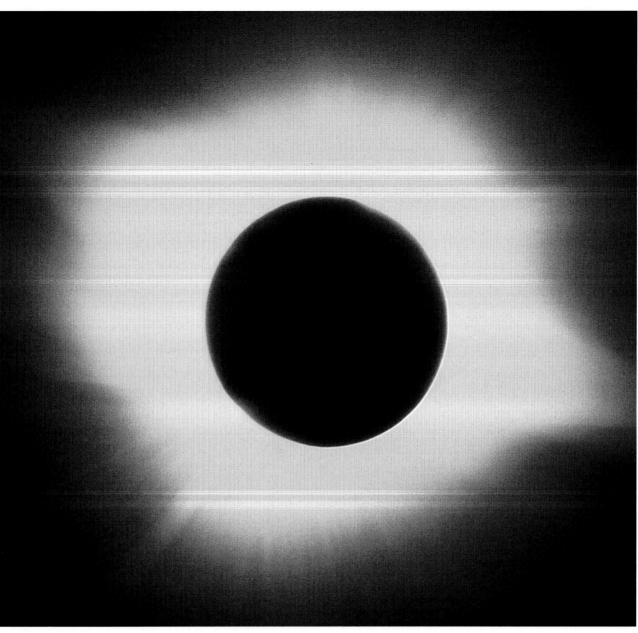

CORONA. During a total solar eclipse, the corona—the Sun's faint, wispy outer atmosphere—becomes visible. Temperatures in the corona exceed 1 million degrees, but the gas is so tenuous that you could stick your hand in a jar full of coronal material without feeling anything.

A STAR IS BORN

To borrow the words of the Grateful Dead, star formation is a long, strange trip. It starts with a dust grain the size of a particle in cigarette smoke, a particle floating aimlessly in the dark and frozen void of deep interstellar space. The process culminates millions of years later in the violence of whirling disks, supersonic jets, and furious stellar winds. The Hubble Space Telescope has given astronomers a front-row seat to this celestial fireworks display.

Contrary to popular perception, outer space is not totally empty. Scattered about the immense distances between the stars is an occasional dust particle here, a hydrogen atom there. Over hundreds of millions of years, gravity brings these gas and dust particles together into enormous clouds that span hundreds of light-years. Driving at 60 miles (96 kilometers) per hour, it would take you 3 billion years to travel from one end of a cloud to the other.

These clouds are stellar nurseries, containing enough material to form tens of thousands of stars. Yet they can also be thought of as stellar graveyards, for they contain the ashes of stars that have long since exhausted their nuclear fuel and passed away into the interstellar night.

Astronomers have found about 6,000 of these so-called molecular clouds in our Galaxy. Once again, hydrogen and helium dominate, accounting for roughly 90 percent and 10 percent of the atoms, respectively. But molecular clouds also contain a myriad of complex molecules, including water, alcohol, ammonia, and various carbon-based organic compounds. In other words, molecular clouds contain the raw ingredients for life.

For eons, not much happens inside these clouds. The various gas and dust particles jostle about, but the cloud pretty much stays the same, though it teeters on the edge of gravitational collapse. Sooner or later, a shock wave from a nearby supernova, a stellar wind from a nearby massive star, or a density wave rolls through the cloud, providing just enough impetus to allow the inexorable pull of gravity to take over.

Over the course of a few million years, thousands of relatively dense spinning concentrations of gas called globules form within the cloud. The globules have more mass and thus stronger gravitational forces than the surrounding material, allowing them to attract even more material. When they reach a certain density, they are called dense cores. The dense cores continue to contract under gravity until they become bona fide protostars. Many of these dense cores break into two or more clumps, leading to the formation of binary and multiple star systems.

Protostars don't yet have sufficient heat and pressure in their cores to ignite fusion reactions, but they shine brightly as they convert the energy of gravitational contraction into heat and light. At this stage, protostars are many times larger than the Sun, even if they contain less overall mass.

As gravity pulls a cloud inward, the cloud starts to spin more and more rapidly. Just as a ball of dough spun in the air by a pizza chef forms a flat crust, a rapidly spinning gas cloud eventually flattens into a disk. Friction causes material in these giant maelstroms to spiral inward, feeding more and more material to the ravenous protostar. Prior to Hubble's launch, ground-based telescopes had photographed a handful of such disks.

Hubble has found a veritable gold mine of disks in the Orion Nebula. At a distance of 1,500 light-years, the Orion Nebula is one of the closest large star-forming regions and a familiar sight to amateur astronomers. Hubble, however, reveals the turbulent inner region of the nebula as it has never been seen before. Using Hubble, Rice University astronomer C. Robert O'Dell and his colleagues have spotted material surrounding 153 of Orion's embryonic stars, roughly half the stars surveyed. These observations prove that disks are a natural outcome of starbirth.

Because all the planets in our solar system go around the Sun in the same direction and in approximately the same plane, it's almost certain that they assembled themselves inside a disk similar to the ones in the Orion Nebula. Some of Orion's disks are considerably larger than our solar system and have more than enough material to form planetary systems like our own. "The building blocks are there to

form planets," says O'Dell, "however, it doesn't mean planets will certainly form."

As a star emerges from its gaseous womb, the disk gives rise to spectacular jets of hot gas that shoot across immense volumes of space. Once again, astronomers knew about the jets prior to the launch of Hubble, and they also knew the jets contained dense clumps of material. But ground-based telescopes could not resolve the detailed structure of either the jets or the clumps, nor could they tell whether a star's jet originated near the star or far from it. Because of these uncertainties, astronomers developed a wide range of models to explain these jets.

When Hubble beamed its first jet images to Earth, entire classes of models fell by the wayside. The Space Telescope revealed some jets to be as tightly confined to narrow beams as lasers, with some stretching for tens of light-years. It also showed that jets originate very close to the protostars. Hubble resolved the structures of the clumps, whose shapes proved that the clumps are being shot out in machine-gun fashion rather than originating from interactions inside the jets.

Although astronomers still don't know exactly how the jets form, they think magnetic fields redirect a fraction of the material in the disks and funnel it in a direction perpendicular to the plane of the disk, forming twin jets that stream away in opposite directions along the protostar's spin axis. The specific mechanism that collimates the jets is still unknown, although theorists have put forth several viable models.

A team of astronomers led by Chris Burrows of the Space Telescope Science Institute took two Hubble images 11 months apart of the disk-jet system HH-30. By tracking how far the clumps moved during the 11-month interval, Burrows's team determined that the jet material moves at speeds of 500,000 miles (800,000 kilometers) per hour.

The disk sporadically shot out the clumps in 20- to 50-year intervals. The young Sun almost certainly had a disk and jet too, so HH-30 provides us with a glimpse of what the solar system looked like 4.5 billion years ago.

Since jets originate in disks, the fact that clumps can be seen inside jets suggests that disk material falls onto protostars in spurts. Although astronomers are still debating why this happens, Mario Livio of the Space Telescope Science Institute provides one possible explanation: "The disk builds up, builds up, builds up, and then at some point when it reaches some critical surface density, it becomes unstable. The viscosity increases, and boom, it dumps material onto the star."

Eventually, a protostar gobbles up enough mass from its surrounding disk that the release of gravitational energy heats the core to a temperature of 5 million kelvins. This is hot enough to boost the speed of the positively charged protons to an energy at which they can overcome their repulsive tendencies and start fusing into helium. The nuclear furnace ignites. The outward push of heat and radiation from the core counterbalances the inward pull of gravity. The protostar no longer needs the "proto"; it has officially entered the realm of stardom. The whole process takes a million years for massive stars, perhaps 50 million years for stars like the Sun, and 100 million years for extremely low-mass stars.

Some stars never quite make it. They have so little mass that they never compress their cores to densities that allow fusion to be sustained. These failed stars are known as brown dwarfs. The cut-off point that distinguishes a star from a brown dwarf is about 0.08 solar mass (1/12 the mass of the Sun, or 80 times the mass of Jupiter). Any object above 0.08 solar mass will sustain fusion reactions in its core, and any object below the threshold will not. Because brown dwarfs shine only dimly from the energy of gravitational contraction, they

are difficult to find. Astronomers suspected brown dwarfs were out there, but for decades they searched unsuccessfully.

Finally, in 1994 a team of astronomers from the California Institute of Technology and Johns Hopkins University found a brown dwarf orbiting the red dwarf star Gliese 229, located about 19 light-years from Earth. The team later imaged the brown dwarf, now called Gliese 229B, with Hubble. The Hubble observations prove beyond all doubt that Gliese 229B orbits the star and is not a much larger background object that just happens to lie along the same line of sight. Astronomers will use Hubble over the coming years to track the orbital motions of the star and brown dwarf, which will help them pin down the brown dwarf's exact mass.

Not only has Hubble contributed to the discovery of the first brown dwarf, it also may have found the signature of a planetary system. In 1984 astronomers discovered a dusty disk around the young star Beta Pictoris, located some 50 light-years from Earth. Hubble's ultra-sharp eye has found a warp in the inner disk. Left on its own, the disk should flatten out in just a few thousand years, a blink of the eye on the cosmic time scale. Something, perhaps one or more planets, is maintaining the warp.

A molecular cloud can spawn thousands upon thousands of newborn stars. The new stars repay their debt by engaging in matricide, killing the mother cloud from which they were born. The most massive newborn stars pump out prodigious quantities of ultraviolet radiation, with devastating consequences for their surroundings. Hubble has provided several spectacular images of this destructive process; perhaps the most stunning of all is the famous "elephant trunks" photo of the Eagle Nebula.

The image, taken by Jeff Hester and Paul Scowen, shows three columns of cold hydrogen gas one to three light-years in length.

Stars are forming inside these pillars. Massive young stars, located off the top edge of the image, bombard the pillars with nasty doses of ultraviolet radiation. The radiation slowly erodes the pillars in a process called photoevaporation. "It's like taking a blowtorch and holding it to a block of ice," says Hester. The heated gas speeds away from the cloud as a powerful wind. Hester stresses that only Hubble could see the process of photoevaporation at work. "Even though the pillars themselves are quite large, the scale over which the physics happens, at the surface of the cold, dense cloud, is extremely small. To see what's happening at that interface, you need the resolution of Hubble."

Hester and Scowen have seen the destructive power of photoevaporation at work in the Lagoon Nebula as well. They used Hubble to zoom into the inner bowels of the nebula, a region called the Hourglass. There, a massive young star named Herschel 36 is blasting away at the surrounding nebula with its ultraviolet radiation. The resulting photoevaporative winds interact with Herschel 36's stellar winds to form eerie tornado-shaped structures. Hester points out that photoevaporation shuts down the star-formation process. "It cuts the stars off from the material they were forming from. It forces one to consider the impact of massive stars on truncating the growth of stars that are forming near them."

Massive stars like Herschel 36 aren't just bad news for stars, they're also bad news for planets. At the heart of the Orion Nebula lie four massive stars collectively known as the Trapezium, the hottest of which has a surface temperature of 40,000 kelvins and the luminosity of 250,000 Suns. A team led by Doug Johnstone of the Canadian Institute for Theoretical Astrophysics in Toronto has analyzed Hubble images of disks situated within one light-year of the Trapezium. Ultraviolet radiation from the Trapezium stars is eating away at the disks, destroying them from the outside in. "It's kind of sad that what we're seeing here is quite likely the destruction of protoplanetary disks, disks that may very well *not* form planets," says Johnstone.

Eventually, the combined radiation and wind from the hundreds or thousands of young stars completely destroys the mother cloud. These stars remain gravitationally bound together in congregations known as open star clusters. The Pleiades, or Seven Sisters, is the most well-known open cluster. Eventually, the stars disperse and the cluster breaks apart. Each star is then free to orbit the center of the Galaxy on its own. What sort of life lies ahead for these stellar babes in the wood?

Phase 1
As a large gas cloud begins
contracting because of gravity, dense
clumps form within the cloud.

Binary core

Phase 2
Some clumps form
single contracting cores,
while others form binary cores.

Single core

Jet

Phase 3
A spinning core flattens into a disk.
Soon thereafter, magnetic fields
lift material off the disk and funnel
it into two counterflowing jets.

Protostar

Disk

Phase 4
Gas and dust in the disk begin clumping
into protoplanets. The jet turns off. Nuclear
reactions ignite in the star's core.

Phase 5
Planets form inside disk.

Sue Biebuyck

HOW STARS FORM

THE HORSEHEAD NEBULA protrudes from a giant gas cloud 1,600 light-years from Earth. Dust in the nebula blocks light from more distant stars. The "horsehead" is about one light-year tall.

Jerry Lodriguss

STELLAR NURSERY. The great Orion Nebula is a stellar nursery 1,500 light-years from Earth. The nebula spans 11 light-years. You can see this nebula with your naked eye—look for the hazy patch of light below the belt in the constellation Orion. The nebula is actually a hot spot on the edge of a giant gas cloud.

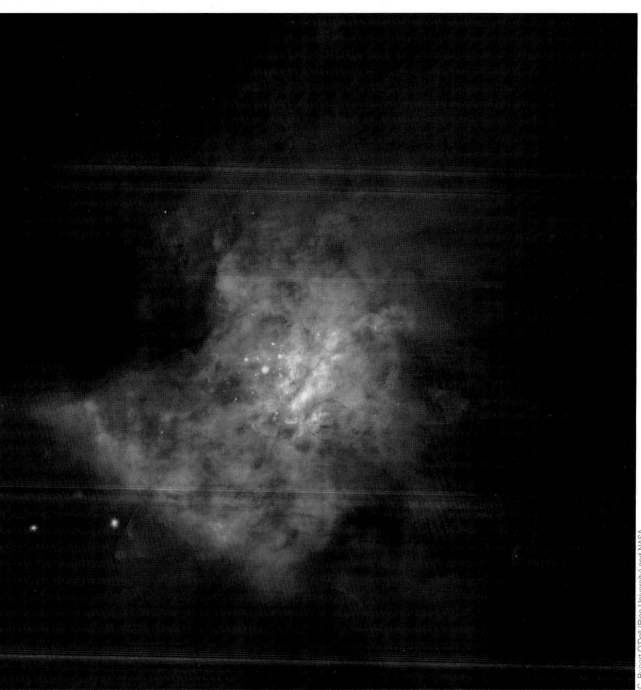

Robert O'Dell (Rice University) and NASA

INNER ORION. Hubble peers deep into the heart of the Orion Nebula, revealing a violent realm of hot stars, shock waves, and jets. Astronomers from Rice University in Houston painstakingly assembled this mosaic from 200 separate Hubble images. The distance from top to bottom covers 2.5 light-years. The four bright pink stars in the center, called the Trapezium, are massive young stars much hotter and more luminous than the Sun. Collectively, the Trapezium stars pump out huge quantities of ultraviolet light, which heats the surrounding gas and causes it to glow. Unlike most Hubble photos, the color in this image is close to what the human eye would see if an observer could view the nebula up close. This image covers a patch of sky about 5 percent the size of the full Moon.

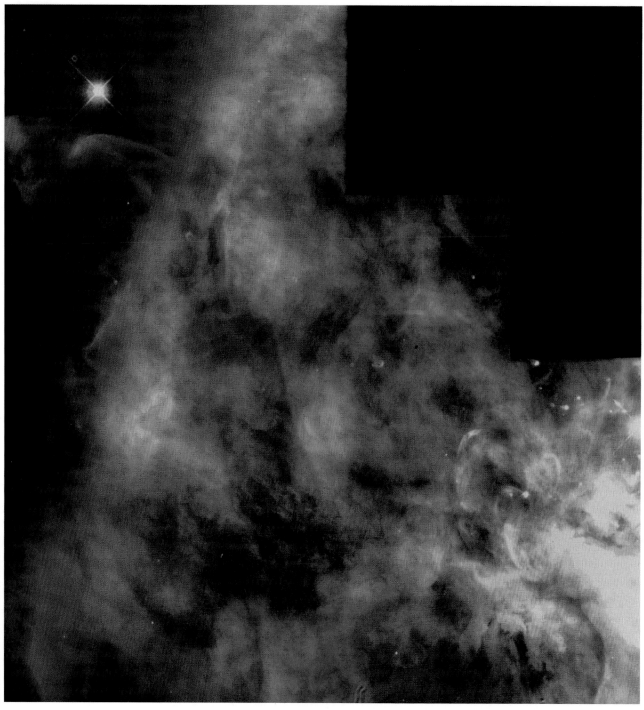

DISKS AND PLUMES. Hubble views a small southeastern portion of the Orion Nebula. Many of the young stars are surrounded by disks of gas and dust, disks that may later form planets. The giant plume in the upper left corner is a giant shock wave generated as a jet of gas plows through the surrounding gas. The jet originates from one of the young stars. The distance from the upper left corner to the lower right is 1.6 light-years.

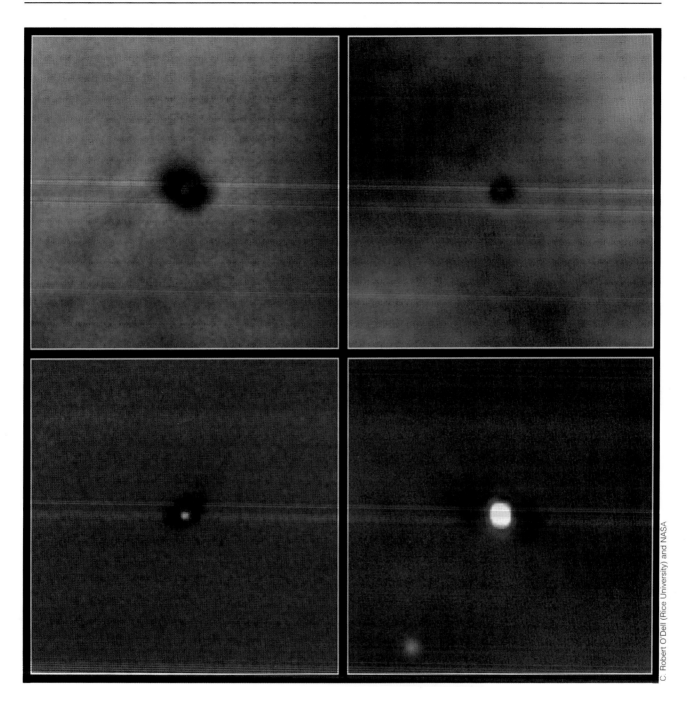

SOLAR SYSTEMS IN THE MAKING? Hubble's remarkable resolution enabled it to view material orbiting at least 153 protostars in the Orion Nebula, about half the total surveyed. Here we see four disks, which look dark because they are silhouetted against the bright backdrop of hot gas; they are 99 percent gas and 1 percent dust. The protostars are the red dots in the center. The disks range in size from two to eight times the diameter of the solar system. If planets have already formed, they are too small and dim to be seen, even by Hubble.

FEARSOME FOUR-SOME. Hubble zooms in on the Trapezium. Ultraviolet light from the brightest of the four Trapezium stars, Theta[1] C Orionis, is so intense that it pushes material away from it, forming the tail-like structures streaming away from nearby stars. The distance from side to side is only half a light-year.

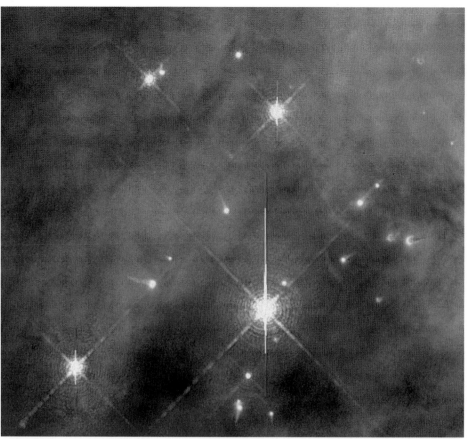

John Bally, Dave Devine (University of Colorado), Ralph Sutherland (Australian National University), and NASA

ZAPPED. The Trapezium stars bathe their stellar neighbors in nasty ultraviolet radiation. Many of these smaller stars are surrounded by disks of gas and dust similar to the one that formed the planets in our solar system. Right before our eyes, Hubble shows ultraviolet radiation evaporating material in the disks. This process might strip enough material from the disks to prevent large planets from forming.

John Bally, Dave Devine (University of Colorado), Ralph Sutherland (Australian National University), and NASA

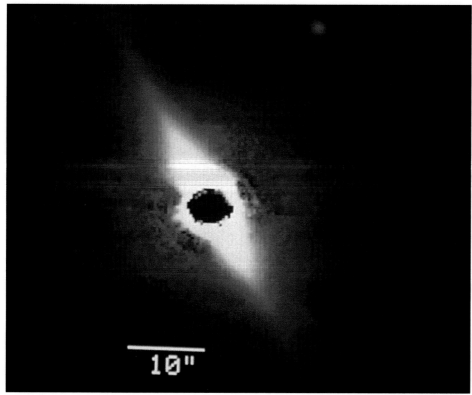

DUSTY DISK. In 1984 astronomers discovered a disk of dust orbiting Beta Pictoris, a star 50 light-years from Earth. Beta Pictoris is a young star, although it is several million years older than stars in the Orion Nebula. Almost all the gas originally in the disk has been swept away by Beta Pictoris's stellar winds. It's possible that planets have already formed inside this disk. In this ground-based image, taken in Hawaii, a specially constructed shield blocks the light from the star.

Paul Kalas and David Jewitt (University of Hawaii)

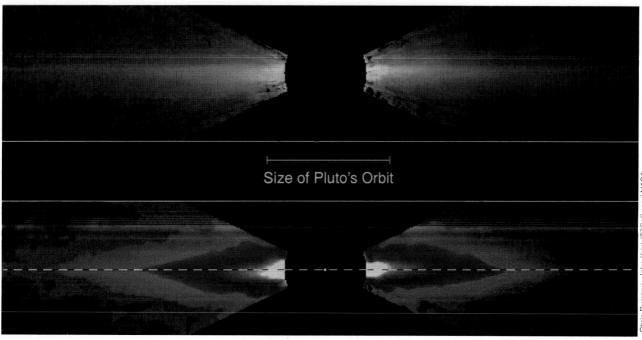

Size of Pluto's Orbit

Chris Burrows, John Krist (STScI), and NASA

POSSIBLE PLANET. Astronomers used this extreme false-color Hubble image (bottom) to discern the structure of the disk orbiting Beta Pictoris. If you look closely, you can tell that the inner region, colored in pink and white, is slightly tilted with respect to the outer disk. Left on its own, the disk would flatten out within a few thousand years, a blink of the eye on a cosmic time scale. A planet roughly analogous to Jupiter could be providing the gravitational force necessary to sustain the warp.

JETS AND DISKS GALORE. Stars rarely form in isolation. Instead, they form in gigantic stellar nurseries with hundreds or thousands of neighbors, many of which have their own disks and jets. The disks feed new material to the growing protostars, while magnetic fields within the disks give rise to the jets.

Jeff Hester (Arizona State University), et al., and NASA

SHOOTING STAR. Hubble has given astronomers their first detailed look at jets of hot gas shooting away in opposite directions from newly forming stars. Most of these protostars, including the one in this image, shoot twin jets in opposite directions along their polar axes. In the top image, the protostar is hidden in the center behind a thick cloud of gas and dust from which it is forming. Each jet spans about half a light-year. The bright spots at the far left and right are the regions where the jets have slammed into the surrounding interstellar gas at 500,000 miles (800,000 kilometers) per hour. The sudden deceleration lights up the gas and forms a bow shock pattern similar to the one that forms in front of a speed boat. The two bottom images are details from the top image.

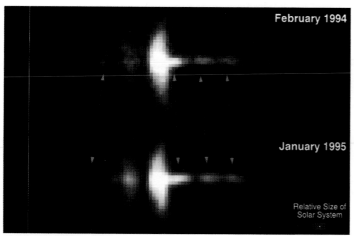

February 1994

January 1995

Relative Size of
Solar System

Chris Burrows (STScI and ESA), et al., and NASA

CLOCKING THE JETS. The HH-30 disk-jet system appears edge-on. Astronomers used Hubble to take two images of the system 11 months apart. For reasons not yet completely understood, many jets contain clumps of gas. By tracking the motion of the clumps with Hubble, the team measured a speed of 500,000 miles (800,000 kilometers) per hour. The images also show that the jet originates very close to the protostar.

MACHINE GUN. Scientists believe that young stellar jets originate in the disks. Magnetic fields channel material off the disk and funnel it along the polar axis of the protostar. For some unknown reason, many disks shoot out clumps of matter every 20 to 50 years. Because disks feed material to the growing protostars, the presence of clumps indicates that stars accrete matter in a sporadic fashion. The green arrowhead in this Hubble image is the classic bow shock pattern that forms when jet material slams into stationary interstellar gas.

Jeff Hester (Arizona State University), et al., and NASA

LONG JET. This Hubble image shows just a small inner portion of a jet that stretches across more than 30 light-years of space. This is one of the longest jets observed to date. It would span more than seven times the distance to Proxima Centauri, the Sun's closest stellar neighbor. Astronomers are trying to figure out how this jet and others remain so tightly collimated over such vast interstellar distances. The jets seen emanating from young stars probably form in the same way as jets that shoot away from black holes.

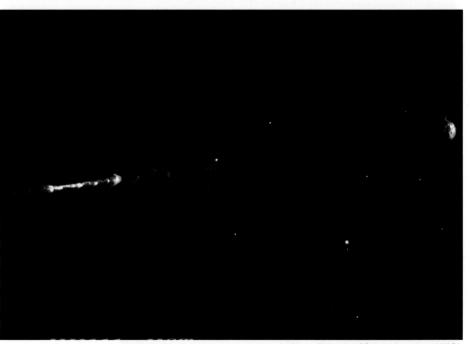

Jon Morse (University of Colorado), et al., and NASA

WIGGLING JET. In this Hubble image a newborn star is shooting out a jet 3 trillion miles into space, equivalent to half a light-year. The newborn star itself is invisible; it's embedded in its nebulous stellar nursery at the lower right. The star may have a binary companion whose gravity causes the star's spin axis to precess like a wobbling top. This motion may give the jet some of its wiggling appearance. A counterjet going in the opposite direction is obscured by the dense, dark cloud.

THE EAGLE NEBULA is a stellar nursery located 7,000 light-years from Earth. The nebula's gas is so hot that electrons are stripped from their atoms. Massive young stars illuminate and heat the nebula. Ground-based photos such as this show the nebula in its entirety, whereas Hubble zooms in on the detailed structures. The Space Telescope thus complements rather than supplants ground-based telescopes.

Jeff Hester, Paul Scowen (Arizona State University), and NASA

PILLARS OF CREATION. Hubble zeroes in on the three "elephant trunks" of the Eagle Nebula. Stars are forming inside the pillars, which are columns of cold, dense hydrogen gas. The left pillar is 3 light-years long. Massive newborn stars above the top edge of the image illuminate the pillars, providing the spectacular sense of depth. These hot stars pump out torrents of nasty ultraviolet radiation, which slam into the pillars, eroding them in a process called photoevaporation. Neither this image nor the ground-based image on the facing page shows what the nebula's colors would look like to the human eye.

COSMIC EGGS. Here we see the top of the left pillar. The tiny fingers are about the size of our solar system. Ultraviolet light from nearby stars photoevaporates gas off the surfaces of the pillars, forming the glow at the top of the pillar. As the pillar boils away, even denser pockets of gas emerge known as "evaporating gaseous globules," or EGGs for short. The EGGs are appropriately named because embryonic stars are forming within them. Some of the EGGs in this image are at the tip of the fingers; others have broken off and appear as teardrop-shaped structures. Photoevaporation will eventually destroy the EGGs as well, inhibiting the growth of the future stars by limiting the amount of material available to them.

Michael Stecker

THE LAGOON NEBULA is a stellar nursery located about 5,000 light-years from Earth. Ultraviolet radiation from hot, young stars embedded in the 120-light-year-wide nebula strips electrons from their host atoms. The electrons whiz around the nebula at near light speed until they crash into and mate with other atomic nuclei. The resulting atoms emit red light, giving the nebula its vivid photographic color. The same process is responsible for the Eagle Nebula's red color. Hubble has imaged the boxed area at the center of the nebula.

Adeline Caulet (ST-ECF and ESA), and NASA

THE HEART OF THE LAGOON. Hubble zooms in on the Lagoon's inner regions. The Hourglass lies at the upper left, with the bright star Herschel 36 right next door. Ultraviolet radiation from Herschel 36 slams into the surface of the surrounding nebula, heating the cold gas to temperatures hotter than the surface of the Sun. "It's like taking a blowtorch and holding it to a block of ice," says astronomer Jeff Hester. This is photo-evaporation at work, the same process that is destroying the Eagle Nebula's elephant trunks.

THE LAGOON'S HOURGLASS. Hubble reveals the Lagoon's Hourglass as it's never been seen before. Heat and ultraviolet radiation from Herschel 36 generate tremendous disturbances in the nebula, creating powerful photoevaporative winds that sweep material out of the Hourglass. The winds interact with a furious wind of charged particles that also emanates from Herschel 36. This complex interaction sculpts the eerie tornado-shaped structures, the largest of which is a mere one-half light-year long.

JUST MADE IT. Hubble catches one of the least luminous stars known to astronomers, a star that falls just above the mass threshold needed to ignite nuclear fusion. The star appears as a barely discernible blip located at 4 o'clock from its brighter and more massive stellar companion. It has only one-tenth the Sun's mass and $1/60{,}000$ the Sun's luminosity. If the star were as far away as the Sun, it would only be eight times brighter than a full Moon. This binary system lies 25 light-years away. The two stars are separated by twice the Earth-Sun distance, too close for the smaller star to be seen in ground-based telescope images.

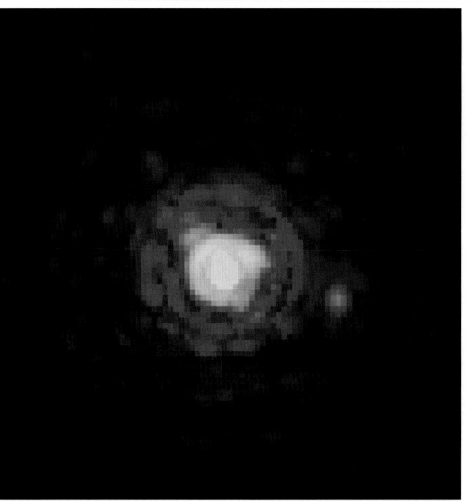

Cesare Barbieri (University of Padua), et al., and NASA/ESA

GLIESE 229B. The little dot near the bottom of this Hubble image is a big deal for astronomers. It is the first confirmed brown dwarf—an object that forms like a star but does not have enough mass to sustain nuclear reactions in its core. The brown dwarf orbits the star Gliese 229 (the much brighter star in the center) at a distance comparable to Pluto's distance from the Sun. The system as a whole is 19 light-years from Earth. The brown dwarf's surface temperature is about 1,000 kelvins, about 1,000 kelvins cooler than the coolest star.

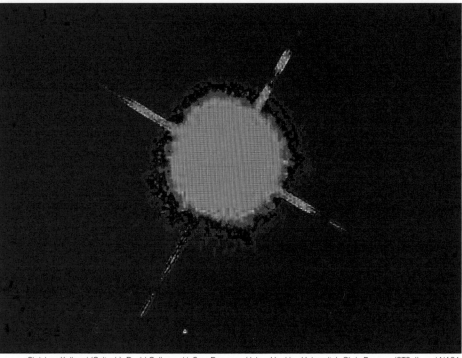

Shrinivas Kulkarni (Caltech), David Golimowski, Sam Durrance (Johns Hopkins University), Chris Burrows (STScI), and NASA

Michael Carroll

BROWN DWARFS like Gliese 229B are actually red. Gliese 229B shines feebly as it converts gravitational energy into heat. Although Gliese 229B is probably 40 to 55 times more massive than Jupiter, gravity compresses it into a sphere the same size as the planet. Essentially stars that look like planets, brown dwarfs share a similar chemical composition with giant planets such as Jupiter.

THE SEVEN SISTERS OF THE PLEIADES. Stars emerging from their stellar nurseries remain gravitationally bound together in groupings called open clusters. The Pleiades, located 360 light-years away, is the most famous open cluster. The brightest of its 400 to 500 stars can be easily seen with the naked eye. Surrounding the cluster is gas that either hasn't been incorporated by stars or was ejected in vast stellar winds. This photo was taken from the ground.

STELLAR SHROUDS

In the 18th century, astronomers looked to the night sky with their small, though ever-improving, telescopes. They started seeing extended, fuzzy disks that looked like the outer planets. William Herschel, the discoverer of Uranus, started calling them "planetary nebulae." The name has stuck ever since.

The name is a misnomer. Planetary nebulae have as much to do with planets as eggplants have to do with eggs. Astronomers now know that planetary nebulae are the fossil remains of intermediate-mass stars like the Sun. If disks and jets offer us a glimpse into the Sun's past, planetary nebulae allow us a sneak preview of what the Sun will look like 5 billion years in the future.

As explained in Chapter 2, stars like the Sun produce most of their energy by fusing hydrogen to helium. The energy from this reaction generates a gas pressure that flows outward, counterbalancing the inward pull of gravity. As long as the star is generating energy in its core, it remains perfectly stable and happy. But stars are born with a finite supply of fuel. When the star runs out of gas, so to speak, the star is headed on the road to ruin.

Scientists who study human behavior vociferously debate the relative importance of nature versus nurture, whether a person's character and personality are more influenced by the genes he or she inherits from his or her parents, or by the culture and environment in which he or she grows up. When it comes to stars, there is no such debate. For a star, it's the genes—the amount of mass the star is endowed with at birth—that determines its luminosity, its temperature, its lifespan, and its death. For stars, mass is just about everything.

The lowest-mass stars can barely sustain fusion in their cores. Compared to the Sun, these stars convert hydrogen to helium at a snail's pace. As a consequence, they give off a feeble, reddish glow. The Sun's closest stellar neighbor—Proxima Centauri—is an example of such a low-mass red dwarf. Proxima Centauri is only 4.25 light-years away, but with a luminosity only $1/10,000$ that of the Sun, it is too dim to be seen with the naked eye. Being small and inconspicuous has its advantages, however. Red dwarfs burn their fuel so slowly that they live almost forever—more than 1 trillion years. Because the universe is "only" 10 to 20 billion years old, every red dwarf ever born is still plodding along with a long lifetime ahead of it.

The high-mass stars, those with more than 10 solar masses, dominate their regions of space with dazzling blue-white light and furious stellar winds. The stars in the Trapezium, for example, are at least 10,000 times more luminous than the Sun. But these stars pay a hefty price for their lofty status. They consume their fuel so voraciously that their lives can be measured in millions of years, the lifetime of a gnat when compared to the age of the universe. These heavyweights can be likened to the Egyptian Pharaohs who enjoyed a brief but resplendent reign. As we'll see in the next chapter, these stars end their lives in a blaze of glory.

Intermediate-mass stars like the Sun shine a respectably bright yellow to yellow-white. They live longer than massive stars, but not as long as red dwarfs. The Sun's lifespan is 10 billion years; clearly many solar-mass stars have perished since the birth of the universe.

As University of Illinois astronomer James Kaler writes, "Something so large and powerful as a star is not going to go quietly." The trouble starts when a star's core runs out of hydrogen. But even after the hydrogen runs out, the star still puts up a good fight for a while before gravity ultimately triumphs.

When the core has converted all of its hydrogen into helium, fusion comes to a screeching halt. Gravity causes the core to contract. This contraction produces heat, which ignites fusion in a shell of hydrogen just outside the core. Heat emanating from the contracting core and the hydrogen-burning shell causes the star's outer envelope to swell. The star grows in size ten- to a hundred-fold. The outer layers cool to a temperature of 3,000 to 4,000 kelvins—the star has become a red giant. The outermost layers are very loosely bound by gravity, so the star starts shedding tremendous amounts of mass. The gentle breeze that was once the stellar wind now becomes a raging hurricane.

The shrinking core eventually becomes hot enough (100 million kelvins) to begin fusing helium into heavier elements, particularly carbon. The outer layers react by contracting and heating up a little bit, giving the star an orangish hue. The bright orangish stars Arcturus and Aldebaran are examples of stars in this phase of the life cycle.

Burning helium gives the star a new but relatively short-lived lease on life. Whereas the core fuses hydrogen for billions of years, it uses up its helium fuel in a few hundreds of millions of years. When the helium runs out, the core contracts again, and the outer layers expand even farther than before. The star becomes an even larger red giant, and once more its outer layers drift off into space. The Sun will expand past the orbits of Mercury and Venus and possibly beyond the orbit of Earth, engulfing the two inner planets. Although all life will have long been extinguished, Earth itself will probably survive as a scorched, airless, molten hunk of rock because the Sun will have lost about half its mass, weakening its gravity and allowing Earth's orbit to expand.

In its red giant phase a star gets to enjoy a fleeting "15 minutes of fame" right before meeting its doom. Because the star has lost so much mass during this phase, the energy released by continued gravitational contraction can't heat the core to a temperature at which it can fuse carbon into still heavier elements. The core slowly contracts under gravity, eventually forming a stellar cinder known as a white dwarf.

White dwarfs are roughly the size of Earth, with masses ranging from about 0.4 to 1.4 solar masses. These stars cram so much mass into such a small volume of space that a teaspoon of white dwarf material would weigh a ton. To find out how much you would weigh on a white dwarf, take your current weight and multiply it by 8,500. Newly minted white dwarfs are among the hottest stars in the universe, with surface temperatures that can exceed 200,000 kelvins. (Recall that the Sun's surface temperature is 5,800 kelvins.) But with no internal energy source, a solitary white dwarf has nothing to do other than slowly cool and fade.

The outer layers shed by the star in the red giant phases ultimately form planetary nebulae. These objects assume a wonderful variety of shapes and are among the most beautiful objects in the universe. They alone could quickly fill up a Hubble's Greatest Hits photo album.

Based on the handful of planetaries that are close enough to show detail in ground-based telescope images, astronomers had developed an elegant and relatively simple model that explained the shapes they were seeing. But with Hubble, astronomers see the nebulae as they've never been seen before. Hubble has imaged a bewildering variety of complex shapes, showing that the formation of planetary nebulae is more complicated than astronomers had thought. "Hubble has changed everything. We're able to see an enormous amount of detail and we're able to find nebulae that are farther away. That's nice for me because planetary nebulae were getting a little bit dull," says University of Rochester astrophysicist Adam Frank, who models the formation of planetary nebulae on supercomputers.

Under the standard model of planetary nebula formation, the dying red giant sheds a slow wind of gas preferentially along its equator, giving the wind a doughnut shape. "Slow" in this instance is relative—the wind speeds hit 55,000 miles (90,000 kilometers) per hour. Thousands of years later, after the star has lost all of its outer envelope on its way to becoming a white dwarf, it gives off a tenuous but fast wind in all directions at speeds topping 10 million miles (16 million kilometers) per hour. Even though the slow wind has a 10,000-year-or-more head start, the fast wind will sooner or later catch up.

That's when the fireworks begin. The fast wind sweeps up the gas in the slow wind. Ultraviolet radiation from the dying star lights up the swept-up gas and voilà, a beautiful planetary nebula graces the sky. But in the directions above and below the slow wind's doughnut, the fast wind continues virtually unimpeded. The result is what astronomers call a bipolar nebula, an hourglass-shaped nebula with two roundish lobes pinched off at the waist.

But many planetaries do not conform to this model of interacting stellar winds in its simplest form, stimulating the creativity of theorists to conceive new models to explain these deviants. One of the most famous "weirdos" imaged by Hubble, the Cat's Eye Nebula, seems to be point-symmetric around the central star. Mario Livio explains what this means: "If you take any feature anywhere on one side of the nebula, and if you were to pass a line from that point through the central star, you would find exactly the same thing on the other side. It's amazing how point-symmetric it is."

Livio and his colleague James Pringle of Cambridge University have developed a model that explains the origin of point-symmetry. The star belongs to a binary system. When one star expands into a red giant, the companion star lies within the bloated envelope. This causes the companion to fall inward and disintegrate, forming a disk. As is true of protostars, where you find a disk you're likely to find a jet as well. The central star bombards the disk with radiation, causing the disk to warp. This in turn causes the jet to wobble. The precessing jet forms a point-symmetric nebula. As Livio explains: "Take a pen and hold it in the middle. Now precess it like a wobbling top. You will automatically produce something that is point-symmetric, because everything that you produce on one side you will also produce exactly on the opposite side."

Jet Propulsion Laboratory astronomers John Trauger and Raghvendra Sahai have imaged a number of planetary nebulae with Hubble. Some of their planetaries conform to the general model, such as their famous orangish Hourglass Nebula (with its sinister-looking "eye in space"). But others clearly do not. Most recently, Trauger and Sahai have imaged several very young planetary nebulae, ones that are so small they show up as indistinct blobs in ground-based telescopes. Hubble shows beyond all doubt that many young planetary nebulae assume complex, "multipolar" shapes at very early stages in their development.

While their conclusions are still preliminary, Trauger and Sahai think multiple jets are turning on and off in various directions, carving out cavities in which the fast wind can expand more easily. The existence of companion objects triggers the formation of these jets. Because of the variety of complex shapes, Trauger and Sahai suggest that many of these companions are substellar, either brown dwarfs or planets. "The name planetary nebula maybe isn't such a misnomer after all," says Sahai. "Maybe there is a deep connection with planets that we hadn't realized."

Planetary nebulae are certainly interesting objects in their own right, but Sahai says astronomers mostly study them to learn what the star was doing before it died: "By studying the structure of the nebula, we can understand the history of mass loss. This is crucial to understanding the evolution of the star itself because the mass loss is so huge that it dominates the evolution of the star in the late stages of its life." He adds, "We've made qualitative leaps in our understanding of what is going on because of the resolution that Hubble is giving us. Hubble keeps giving us new puzzles and that keeps our life exciting."

Hubble has also proved extremely useful in the study of novae, which occur in binary star systems consisting of a white dwarf and a stellar companion. Hydrogen gas from the companion spills onto the white dwarf. The hydrogen slowly accumulates and heats up over the course of a few thousand or tens of thousands of years. Suddenly, the hydrogen layer becomes so hot and dense that it undergoes nuclear fusion. The result is a spectacular explosion as the hydrogen is quickly converted into helium and blasted into space. The explosion is similar to that of a hydrogen bomb, only millions of times more powerful. To observers on Earth, the dim star suddenly appears thousands of times brighter for several weeks before slowly fading back to normal.

Astronomers have been using Hubble to monitor the aftermath of a nova that flared in the constellation Cygnus on February 19, 1992. The first Hubble image of Nova Cygni 1992, taken on May 31, 1993, before the telescope was repaired, revealed an expanding, spherical shell and a barlike structure in the center. Astronomers couldn't tell, however, if the bar was real or an artifact of the flawed optics. Over the past four years, the central bar has disappeared, while the shell has become noticeably oval-shaped in the direction perpendicular to the bar.

Mario Livio has developed a model to explain the evolution of this system. When the nova erupted, the two stars were quickly engulfed inside the shell of debris. The orbital motion of the two stars acted like eggbeaters, concentrating the ejected gas along the orbital plane and forming the bar seen in the first Hubble image. As the shell continues to expand and thin out, the white dwarf's fast wind of hot, tenuous gas rams into the shell. Because the shell was less dense in the direction above and below the orbital plane, the fast wind could push it farther ahead, which explains why the shell evolved from a circle into an oval. "Because Hubble has such fantastic resolution, you can resolve this shell a very short time after the explosion took place and follow it, rather than seeing it decades later," says Livio.

Novae are quite spectacular by human standards. But they are mere firecrackers when compared to supernovae, the cataclysmic fate of the universe's stellar heavyweights.

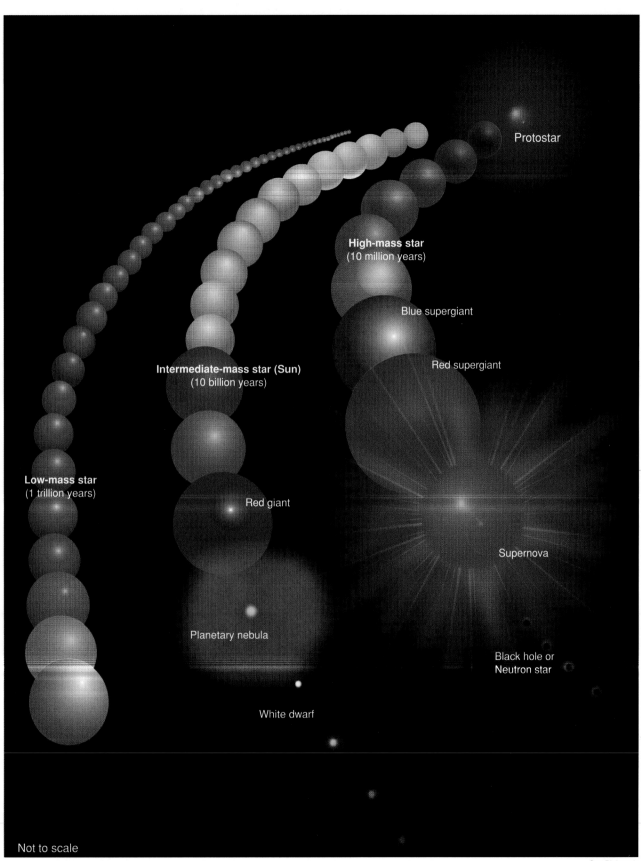

Protostar

High-mass star
(10 million years)

Blue supergiant

Intermediate-mass star (Sun)
(10 billion years)

Red supergiant

Low-mass star
(1 trillion years)

Red giant

Supernova

Planetary nebula

Black hole or
Neutron star

White dwarf

Not to scale

Sue Biebuyck

STELLAR EVOLUTION

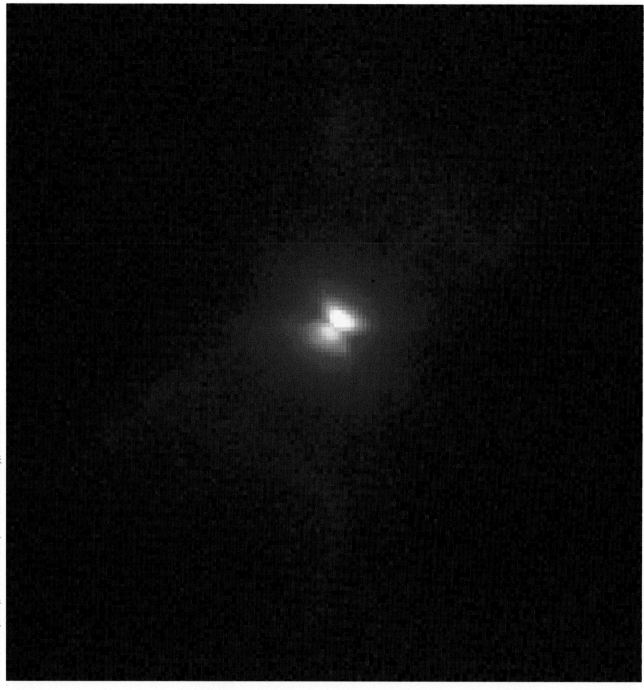

Howard Bond (STScI), Robin Ciardullo (Penn State University), and NASA

RED RECTANGLE. Hubble captures a protoplanetary nebula, the Red Rectangle. Its central star is hidden behind a thick disk of dust, a disk that can't be seen from ground-based telescopes. In this picture, the bright region represents starlight scattering off dust particles in the disk. The star itself is shrinking and heating up on its way to becoming a white dwarf. Previous ground-based observations show that the dying star belongs to a binary system. The binary companion's gravity is probably responsible for forming the disk.

Raghvendra Sahai, John Trauger (Jet Propulsion Laboratory), and NASA

THE EGG NEBULA is a protoplanetary nebula some 3,000 light-years from Earth and 0.6 light-year across. This Hubble image shows the central star enshrouded in a dense cocoon of dust. The star was a red giant just a few hundred years ago and is shrinking into a white dwarf. The nebula's gas and dust were shed by the star during its red giant phase, which lasted several thousand years. The arcs might indicate that mass loss is episodic, varying on time-scales of 100 to 500 years. Starlight appears to be punching through holes in the cocoon to form the searchlight beams.

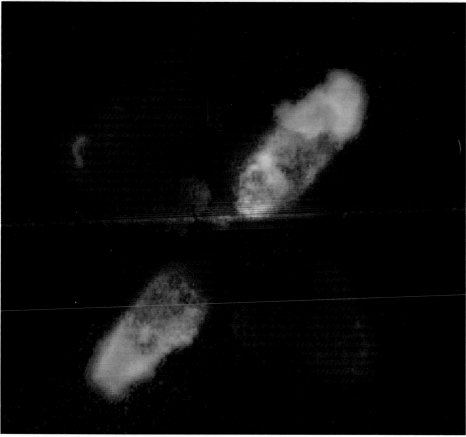

Rodger Thompson (University of Arizona), et al., and NASA

LAST GASPS. Hubble's NICMOS camera—installed by space shuttle astronauts in February 1997—sees infrared light rather than visible light. Because infrared light can pass through the Egg Nebula's thick dust, NICMOS can reveal activity that can't be seen in visible light. NICMOS shows that the dying star is shooting a jet of hot hydrogen gas at 220,000 miles (350,000 kilometers) per hour. When the jet's gas rams into the material in the concentric arcs, the gas heats up, producing the glowing pink tips. The jet spans 200 times the diameter of the solar system. The deep red is heat emitted by hot hydrogen molecules.

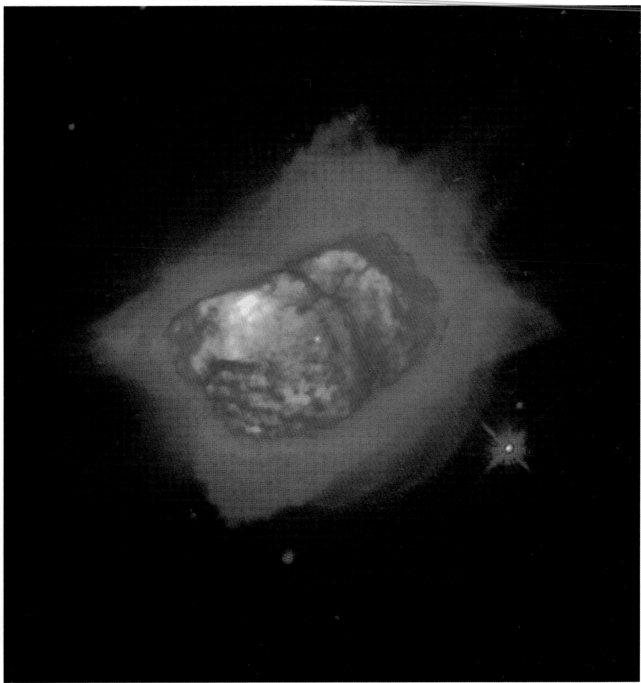

CONCENTRIC SHELLS surround the mysterious young planetary nebula NGC 7027, located 3,000 light-years away. The white dot at the center of this Hubble image is the dying star, one of the hottest known to astronomers. The bright inner regions represent gas violently thrown off by the star right before it died. As is the case with the Egg Nebula, the dimmer outer regions represent gas shed by the red giant star. It's not clear if the concentric arcs represent episodic mass loss or instabilities in the gas.

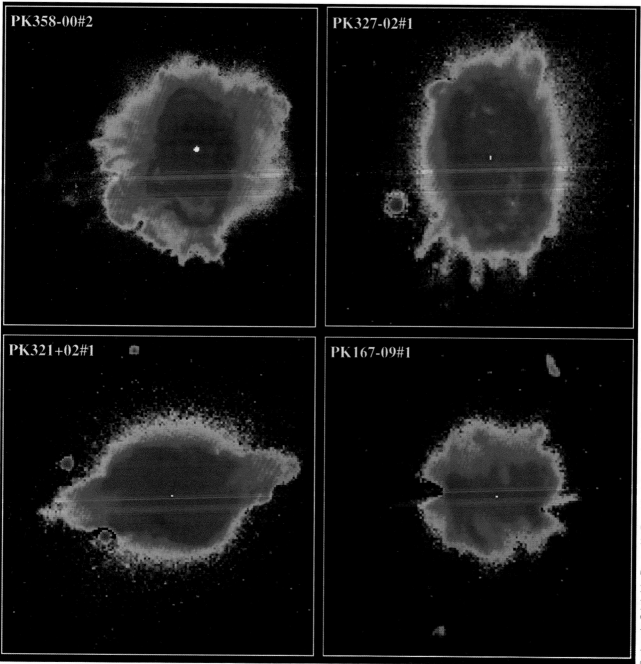

PK358-00#2

PK327-02#1

PK321+02#1

PK167-09#1

Raghvendra Sahai, John Trauger (Jet Propulsion Laboratory), and NASA

YOUNG PLANETARIES. Hubble's ultrasharp resolution allows astronomers to see detailed structure in four planetary nebulae caught in the early stages of formation. These nebulae, which show up as indistinct blobs in ground-based images, display a bewildering variety of shapes. Astronomers think jets of high-speed gas create these complex features. The jets themselves are triggered by companion objects. The wide variety of features suggests that some of the companions are substellar, either brown dwarfs or planets roughly the mass of Jupiter.

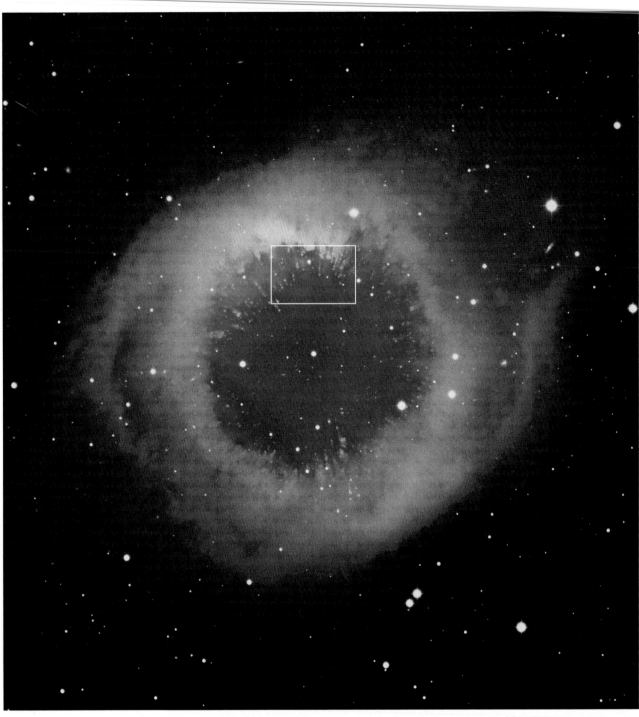

THE HELIX NEBULA is the closest planetary nebula, at a distance of 450 light-years. This ground-based image shows the entire nebula, which is 1.6 light-years across and has an apparent size half that of the full Moon. The white dwarf is located at the center. Although the Helix Nebula is probably an hourglass-shaped nebula, we see the doughnut as a ring because we're looking right down the throat of the nebula. The gas in the bubble above the doughnut is so tenuous that we look right through it. The ring material was ejected by the star more than 10,000 years ago, when it was a red giant. This planetary nebula is much older than the others shown in this book. Hubble imaged the boxed area.

COLLIDING WINDS. Hubble captures remarkable detail in the Helix Nebula. A hot, fast wind of gas from the central star slams into the inner edge of the nebula, breaking apart the slower-moving gas into small, dense globules. The fast-moving wind strips material off these globules, so the tails point away from the central star, forming a radial pattern reminiscent of spokes on a bicycle wheel.

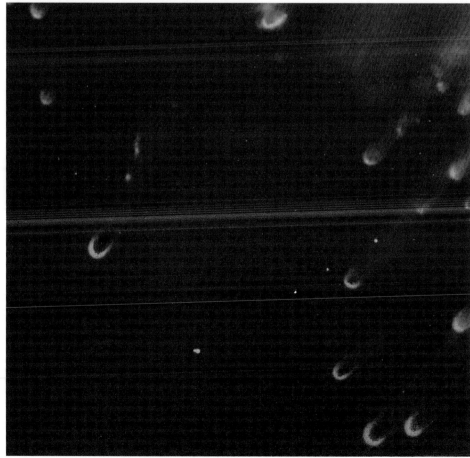

COMETARY KNOTS. Astronomers have dubbed these tadpole-shaped objects in the Helix Nebula "cometary knots" because of their resemblance to comets. In reality, the knots have nothing to do with comets. They are 100 times the Earth-Sun distance across, about twice the diameter of the solar system. The tails stretch across 100 billion miles of space, 1,000 times the Earth-Sun distance. The central star's energetic ultraviolet radiation may eventually erode the cometary knots to nothingness.

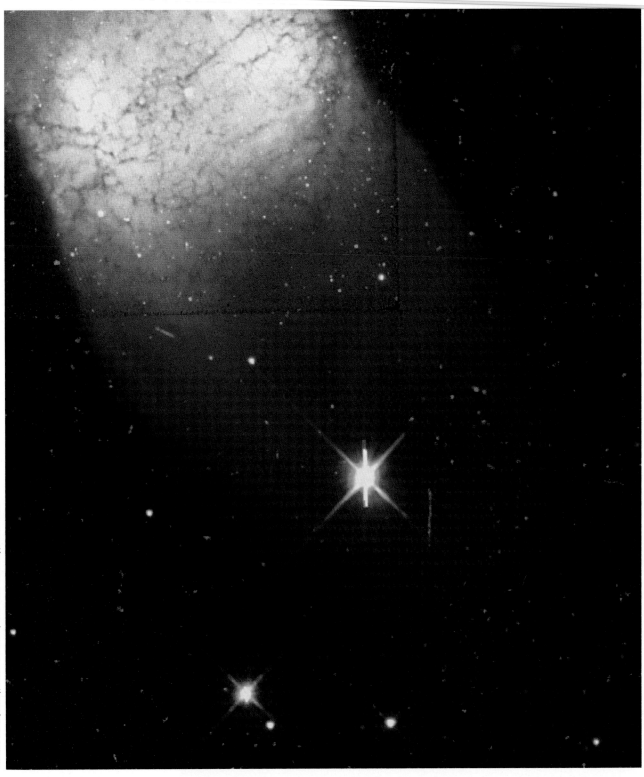

DENSE BELT. Like many planetary nebulae, IC 4406 has a dense belt surrounding the central star's equatorial plane. This is what the Helix Nebula might look like if we could see its doughnut edge-on. Planetary nebulae are often difficult to study because we only see two of the three dimensions. "You could take the same planetary nebula and rotate it, and it will have a completely different appearance," says astronomer Howard Bond.

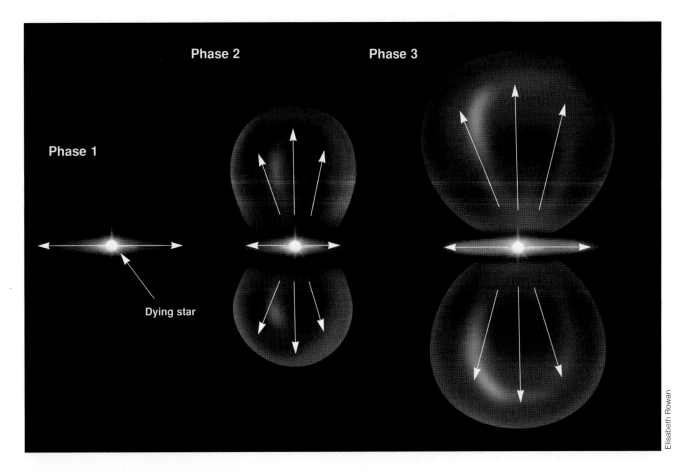

Phase 1

Phase 2

Phase 3

Dying star

Elisabeth Rowan

FORMING AN HOURGLASS. Colliding stellar winds sculpt the classic bipolar hourglass shape of many planetary nebulae. In phase 1, the dying red giant star gives off a slow wind of gas preferentially along its equator, forming a doughnut. In phase 2, the star is shrinking to a white dwarf. The hot star gives off a fast wind that zooms away more or less equally in all directions, quickly catching up to the slow wind that was ejected thousands of years earlier. The material in the doughnut impedes the fast wind's progress, while the fast wind remains unchecked above and below the star's equatorial plane. In phase 3, the planetary nebula assumes a classic hourglass shape.

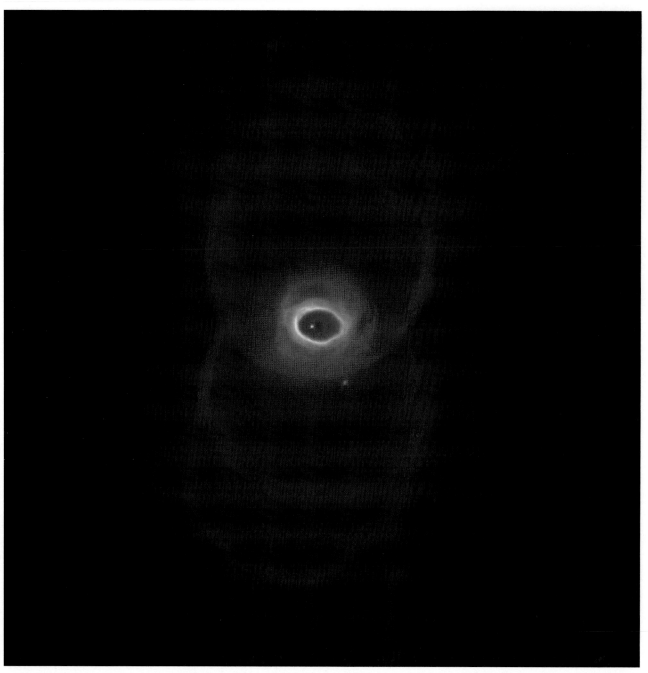

Raghvendra Sahai, John Trauger (Jet Propulsion Laboratory), and NASA

THE HOURGLASS NEBULA, MyCn 18, is a typical bipolar planetary nebula that probably formed when a fast stellar wind rammed into a slow wind ejected thousands of years earlier. Curiously, this Hubble image shows that the white dwarf star is slightly off-center with respect to the nebula. Astronomers suspect that a gravitational interaction with a companion star threw the star from its central position. This nebula is about 8,000 light-years away. Despite the name, this nebula has a completely different origin than the Hourglass region of the Lagoon Nebula.

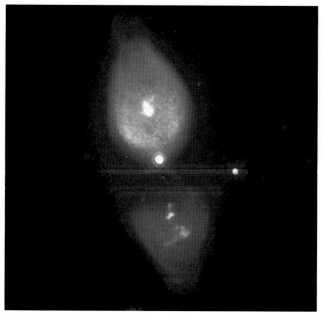

M1-92 is a young planetary nebula that conforms to the common hourglass shape.

HOT STAR. The white dwarf at the center of planetary nebula NGC 2440 has a temperature of 200,000 kelvins, making it one of the hottest known stars. "It's hard to find these stars from the ground because they're sitting in the middle of a bright planetary nebula, but with Hubble it is pretty easy," says astronomer Howard Bond.

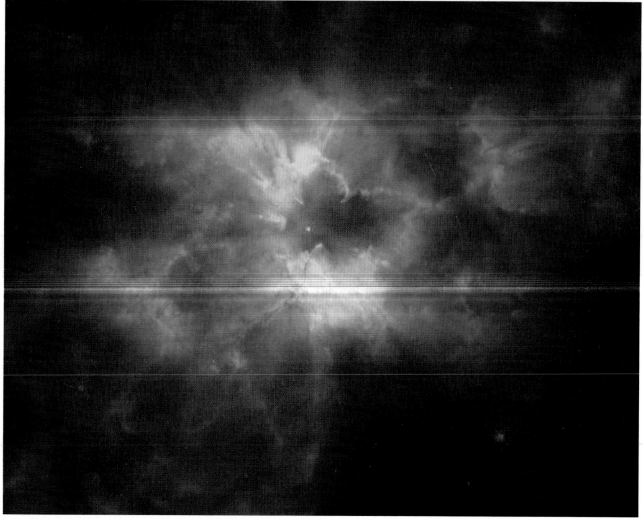

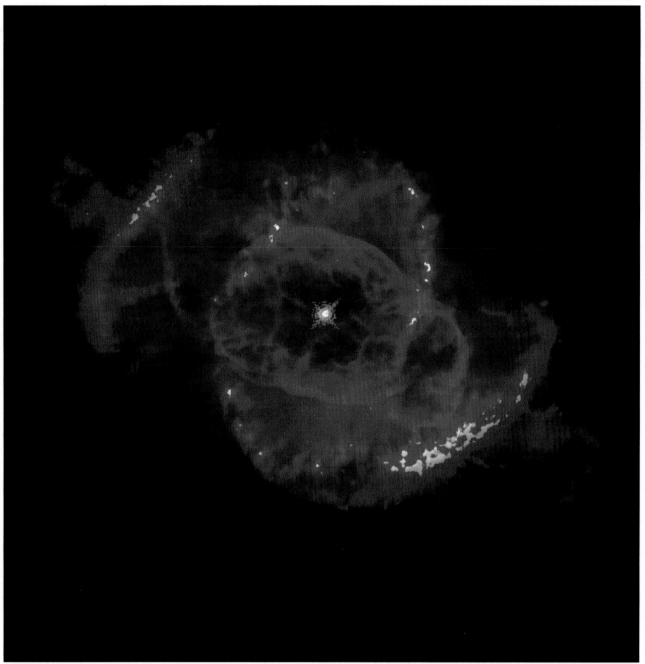

THE CAT'S EYE NEBULA is the archetypical "point-symmetric" planetary nebula—every feature on one side of the central star has an identical twin on the other side. Astronomers think high-speed jets are flowing in opposite directions from the central star, forming identical structures on each side. The jet originates in a disk that contains the remains of a destroyed companion. Radiation from the hot star causes the disk to warp, which in turn causes the jet's axis to precess like a wobbling top.

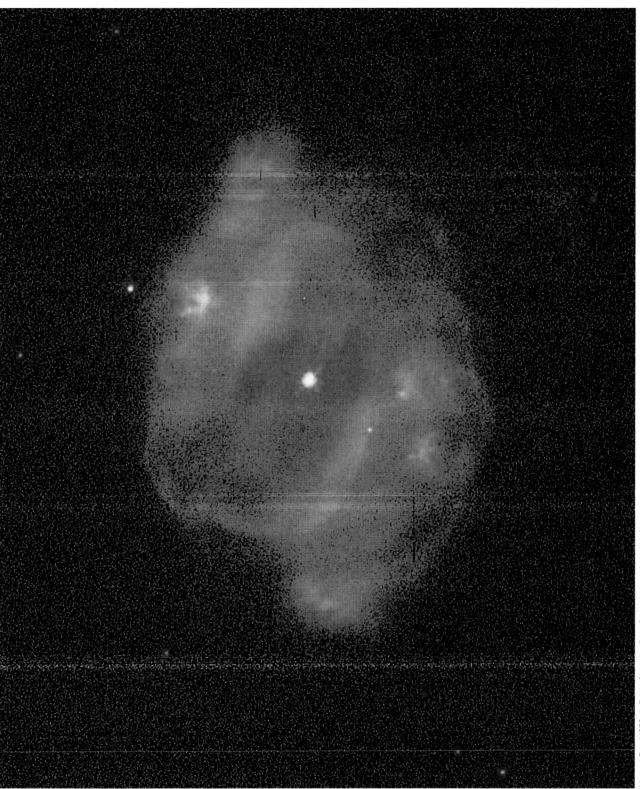

POINT-SYMMETRIC. Hubble captures another point-symmetric planetary nebula, NGC 5307. This nebula appears to be a close relative of the Cat's Eye Nebula.

FLIERs. Hubble is helping astronomers resolve yet another mystery surrounding planetary nebulae: the origin of high-speed jets of gas called FLIERs, seen as red blobs in these images. Many astronomers once thought these blobs were ejected by the star while it was in the red giant phase. From these Hubble images, astronomers identified converging streams of gas above the poles of the central star. The streams appear to be smashing into each other, forming a fast jet of gas. "It's like holding two garden hoses next to each other and letting the streams of water collide. If you do it at the right angle, you get a nice tight spray of water," says team member Adam Frank.

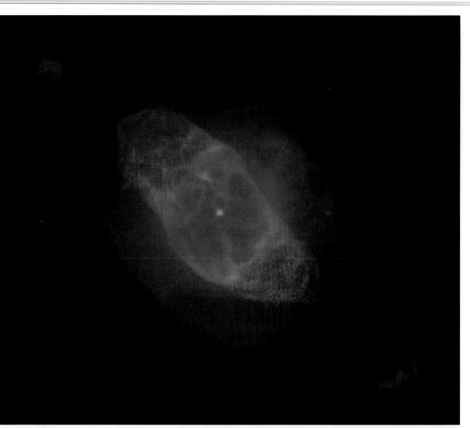

Bruce Balick (University of Washington), et al., and NASA

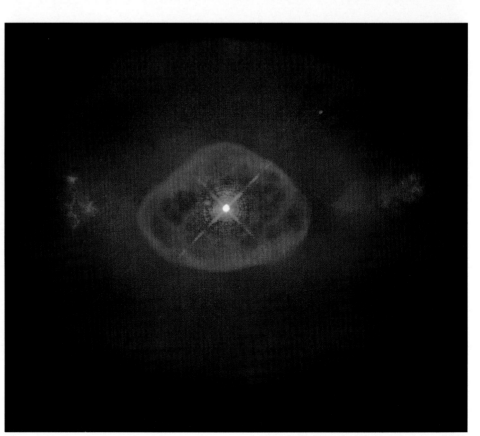

Bruce Balick (University of Washington), et al., and NASA

Michael Shara, Robert Williams (STScI), and NASA

THE "FIREWORK NEBULA." Hubble's sharp eye reveals blobs of gas ejected during a nova that flared in the constellation Perseus in 1901. Astronomers think the firework appearance might be the result of material from the explosion ramming into dense interstellar gas clouds that surround the nova. The collision could be breaking the material up into the glowing blobs seen here. The nova is about 1,500 light-years away. When the nova erupted, it was one of the brightest stars in the night sky. Now it has faded so much that it is 600 times fainter than the faintest naked-eye stars.

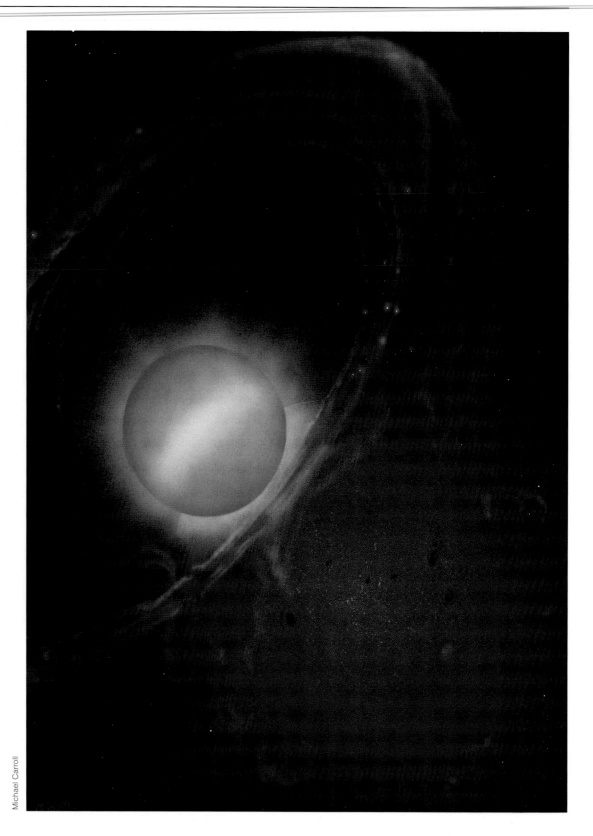

ROAD TO DISASTER. Prior to Nova Cygni 1992, gas from a red dwarf star spilled onto a white dwarf companion. The gas formed a thin disk around the white dwarf, gradually settling onto the surface. When the layer of gas reached a critical temperature, it ignited in a tremendous explosion that launched several Earths' worth of mass into space at speeds topping 700 miles (1,100 kilometers) per second. The debris could cross the Earth-Moon distance in six minutes.

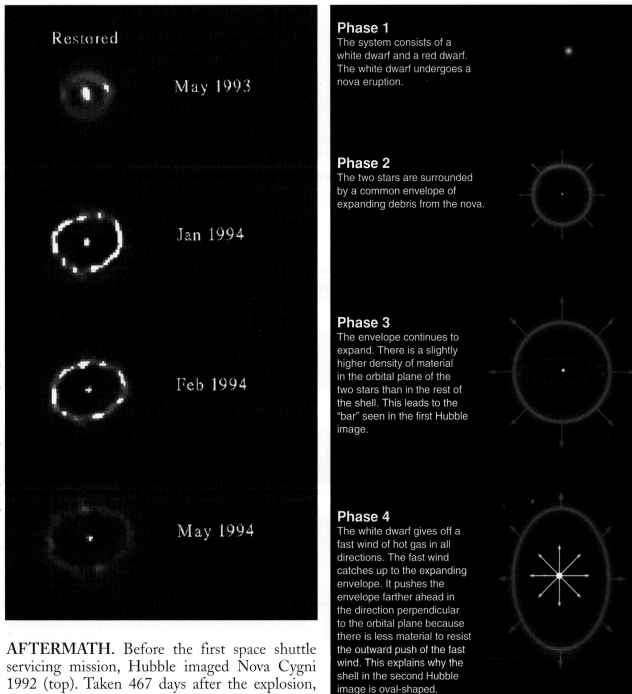

Restored

May 1993

Jan 1994

Feb 1994

May 1994

Francesco Paresce, R. Jedrzejewski (STScI/ESA), Warren Hack (STScI), and NASA

Phase 1
The system consists of a white dwarf and a red dwarf. The white dwarf undergoes a nova eruption.

Phase 2
The two stars are surrounded by a common envelope of expanding debris from the nova.

Phase 3
The envelope continues to expand. There is a slightly higher density of material in the orbital plane of the two stars than in the rest of the shell. This leads to the "bar" seen in the first Hubble image.

Phase 4
The white dwarf gives off a fast wind of hot gas in all directions. The fast wind catches up to the expanding envelope. It pushes the envelope farther ahead in the direction perpendicular to the orbital plane because there is less material to resist the outward push of the fast wind. This explains why the shell in the second Hubble image is oval-shaped.

Sue Biebuyck

AFTERMATH. Before the first space shuttle servicing mission, Hubble imaged Nova Cygni 1992 (top). Taken 467 days after the explosion, the image shows a circular ring and a mysterious bar. The ring is actually the outer edge of an expanding bubble of gas. The bottom three images were taken after the first servicing mission. In these pictures the ring has continued to expand but is no longer circular, and the bar has vanished.

NOVA CYGNI 1992

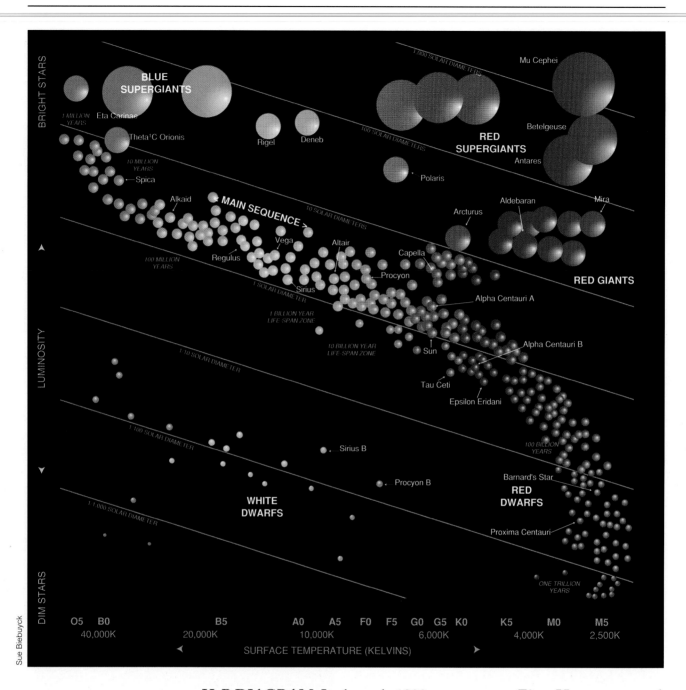

BRIGHT STARS

BLUE
SUPERGIANTS

1 MILLION
YEARS   Eta Carinae

Theta¹C Orionis

10 MILLION
YEARS
←Spica

Alkaid

Rigel   Deneb

Mu Cephei

1,000 SOLAR DIAMETERS

100 SOLAR DIAMETERS

RED
SUPERGIANTS

Betelgeuse

Antares

MAIN SEQUENCE →

100 MILLION
YEARS
Regulus

Vega

Altair

Polaris

10 SOLAR DIAMETERS

Arcturus

Aldebaran

Mira

Capella

RED GIANTS

Procyon

1 SOLAR DIAMETER

Sirius

1 BILLION YEAR
LIFE-SPAN ZONE

Alpha Centauri A

LUMINOSITY

10 BILLION YEAR
LIFE-SPAN ZONE

Sun

Alpha Centauri B

1:10 SOLAR DIAMETER

Tau Ceti

Epsilon Eridani

1:100 SOLAR DIAMETER

Sirius B

100 BILLION
YEARS

Procyon B

Barnard's Star
RED
DWARFS

WHITE
DWARFS

1:1,000 SOLAR DIAMETER

Proxima Centauri

DIM STARS

ONE TRILLION
YEARS

O5   B0   B5   A0   A5   F0   F5   G0   G5   K0   K5   M0   M5
40,000K   20,000K   10,000K   6,000K   4,000K   2,500K

◄   SURFACE TEMPERATURE (KELVINS)   ►

Sue Biebuyck

H-R DIAGRAM. In the early 1900s, astronomers Ejnar Hertzsprung and Henry Norris Russell independently developed a way to classify stars according to temperature and luminosity. The resulting H-R Diagram shows that the vast majority of stars lie along a band called the main sequence, which stretches from the upper left to the lower right. The main sequence is not an evolutionary pathway; instead, it is a snapshot of stellar populations at one moment in time. When stars form, they fall onto a spot on the main sequence depending on mass. The most massive stars lie at the upper left, the least massive on the lower right. Stars spend most of their lives at or near the same spot on the main sequence as they generate energy by fusing hydrogen to helium. When stars run out of their hydrogen fuel supply, they swell in size to become giants and supergiants. High-mass stars explode and leave behind a neutron star or a black hole. Solar-mass stars leave behind white dwarfs, which lie at the bottom of the H-R Diagram.

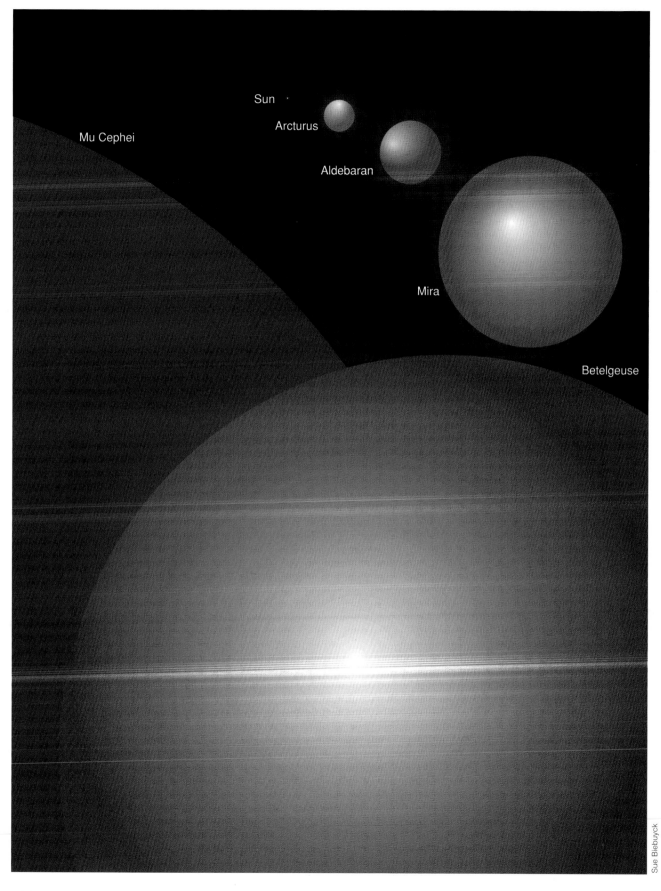

Sun

Arcturus

Mu Cephei

Aldebaran

Mira

Betelgeuse

Sue Biebuyck

STARS LARGER THAN THE SUN

LIFE ZONES. Stars much less massive than the Sun have extremely narrow life zones, the regions where planets can maintain water in a liquid form. The stars constitute the large majority of stars in the Galaxy. Stars more massive than the Sun have wide life zones, but these stars don't live long enough for life to evolve into complex and advanced forms. These stars are relatively uncommon. Stars similar in mass to the Sun offer the best prospects for supporting advanced life.

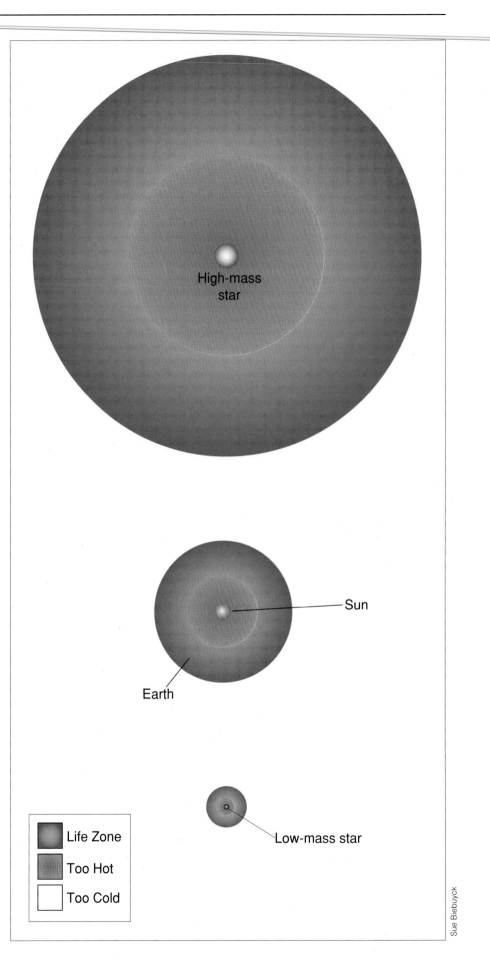

High-mass star

Sun

Earth

Low-mass star

Life Zone

Too Hot

Too Cold

Sue Biebuyck

72

BINARY PLANET. Most stars belong to gravitationally bound systems consisting of two or more stars. This hypothetical planet orbits a binary star in which the two stars happen to be close to each other. Such a planet would very likely have an egg-shaped orbit, leading to wild seasonal climate variations. When the planet is closer to the warming influence of the two stars (right), glaciers melt and water can flow over the surface. Because of the varying climate extremes, the surface of such a planet would not be prime real estate for life.

CATACLYSM

For millions of years, a massive star named Sanduleak −69°202 was the master of its domain. At the height of its powers, the star was so big that 170,000 Suns would have fit inside it. Weighing in at 20 solar masses, the star's nuclear furnace pumped out so much energy that the star shone with the brightness of 20,000 Suns, dominating its region of the Large Magellanic Cloud—a satellite galaxy of the Milky Way Galaxy.

One day, its nuclear furnace ran out of fuel. In a fraction of a second, the core collapsed, triggering a shock wave that ripped through the outer layers, blowing the star to kingdom come. Poor Sanduleak −69°202 was no more.

For 167,000 years, news of the star's catastrophic demise traversed space at the speed of light. On February 23, 1987, it reached Earth, where the explosion was subsequently named Supernova 1987A. Because SN1987A is the brightest supernova since the year 1604, astronomers have dissected every photon of light they could get their hands on to help them better understand the story of stellar death. And Hubble has been at the forefront of this endeavor.

Recall from the previous chapter that when a solar-mass star runs out of helium,

that's the end of the road—the star has insufficient mass to ignite another round of fusion in the core. But a star that started its life with about 10 or more solar masses has sufficient mass at the end of its helium-burning phase to compress its core to a temperature hot enough (1 billion kelvins) to ignite carbon and even heavier elements in succession until the core is made up of iron. At this point, a star's internal structure resembles that of an onion: an iron core is surrounded by a layer of silicon, which in turn is surrounded by shells of increasingly lighter elements.

Iron is the nuclear equivalent of a stop sign. It has the most stable nucleus of any element; it takes more energy to fuse iron than the reaction gives off. So when the core is made of iron, fusion reactions come to a halt and the star is ready to meet its doom.

Gravity takes over and initiates a catastrophic collapse. In less than a second, the inner core collapses. Protons and electrons are squeezed together so tightly that they combine to form neutrons. The resulting stellar cinder is a neutron star, a bizarre object that packs roughly two or three solar masses into a ball the size of a city. To achieve the same density, a giant trash compactor would need to squash Earth into a ball that could fit inside a sports stadium. Neutron stars are so dense that a sugar-cube-sized lump of neutron star material contains as much mass as the entire human population. (A neutron star's gravity is 100 billion times stronger than Earth's. You'd be instantly scrunched into nothingness if you stood on the surface of a neutron star.) Like a lighthouse gone berserk, young neutron stars rotate several dozen times per second. In the case of the most massive stars, the inner core contains more than three solar masses. Gravity squeezes the core beyond the neutron star state all the way down to a black hole.

While the inner core shrinks into a neutron star or black hole, the outer core rebounds with 100 billion g's of acceleration (an F-16 fighter pilot can withstand about 10 g's at most), triggering a shock wave. The breakdown of iron in the inner core produces 10 billion trillion trillion trillion trillion subatomic particles called neutrinos. These ghostly particles could pass through half a light-year of lead as if it weren't there, just as an endless stream of marbles could easily pass through a hula hoop. (Billions of neutrinos are passing through your body as you read this sentence.) Each individual neutrino packs a negligible amount of energy, but there are so many of them that they energize the shock wave from below so it can rip through the entire star and tear it to shreds.

During the few seconds that neutrinos stream out of the collapsing stellar core, the supernova releases more energy in the form of neutrinos than the amount of energy radiated in visible light by all the trillions upon trillions of stars in the universe combined in that same amount of time. In visible light, a supernova can briefly outshine an entire galaxy.

During the explosion, particles and atomic nuclei collide in the stellar envelope, forming the heaviest elements of all, such as copper, gold, lead, and uranium. The explosion casts several solar masses of this precious inventory of heavy elements into deep space, where they provide the raw material for the next generation of stars and planets.

Astronomers have been using Hubble to study Supernova 1987A ever since the telescope was launched in 1990. "Without Hubble, there's no story; we wouldn't be able to study the structure of the supernova in visible light," says Chun-Shing Jason Pun of NASA's Goddard Space Flight Center. "Especially because the supernova is so close, Hubble provides us with a unique chance to study the evolution of a supernova in details we have never dreamed of studying before."

In August 1990 Hubble imaged a 1.4-light-year-wide ring of gas and dust surrounding the expanding debris from the explosion. Then in February 1994, shortly after the first servicing mission, Hubble's sharp eye revealed two larger and fainter rings, each about three light-years in diameter. SN1987A was suddenly a three-ring circus.

Astronomers are still debating the exact origin of the rings. But there is unanimous agreement that all three rings represent gas shed by the progenitor star before it exploded, much as red giants shed their outer envelopes to form planetary nebulae. Astronomers think the star ejected the inner ring material about 10,000 years before it exploded and the outer ring material another 10,000 years before that.

The most recent Hubble images show, for the first time, detailed structure in the expanding debris, which is now about one-sixth of a light-year across and expanding at a speed of nearly 6 million miles (10 million kilometers) per hour. The debris has separated into two distinct blobs that are moving in directions perpendicular to the plane of the inner ring. This geometry strongly suggests that something—perhaps a companion star—has affected both the formation of the ring and the evolution of the expanding debris.

Astronomers are anxiously awaiting the time when the expanding debris collides with the inner ring, an event which should be visible in 2007, give or take a year or two. The ring will suddenly become 1,000 times brighter. With luck Hubble, or Hubble's successor, will be ready to pounce on this event. In the meantime, says Pun, "We will watch the supernova become a supernova remnant."

Astronomers have used Hubble to observe a number of much older supernova remnants. Jeff Hester and Paul Scowen have used the telescope to peer deep into the heart of the Crab Nebula, the tattered remains of a star whose explosive destruction was recorded by Chinese astronomers in the year 1054. Hubble took a series of images of the Crab's inner region over several months. The resulting "movie" revealed a much more dynamic environment than anyone had previously imagined. "In astronomy, normally if you see something change over your career you get excited about it," says Hester. "But the Crab Nebula is so dynamic that if you wait for longer than about a week to get your next look, you'll get the wrong impression because things have changed so terribly much."

The Hubble images show how the central neutron star illuminates the entire 10-light-year-wide nebula, causing it to shine with the energy of 100,000 Suns. The city-sized neutron star rotates 30 times per second. This, coupled with its powerful magnetic field, acts like a slingshot—accelerating a wind of charged particles to near the speed of light. The neutron star's rotation and intense magnetic field are driving two jets of particles that zoom away from the poles at near light speed. Hubble has given astronomers their clearest look at the structure and dynamics of a neutron star jet.

Hester and Scowen are continuing to observe the Crab's inner region with Hubble. "By late 1998 we will have a new Crab movie,"

says Hester, "and whereas the first one was the Crab trailer, this one will be the long-play version."

Hester and Scowen also used Hubble to study the outer fringes of the Crab Nebula. They discovered that the long filaments seen in ground-based images are created by the interaction between a bubble of charged particles bound to the neutron star's powerful magnetic field and the ejecta from the supernova explosion.

Hester has also turned Hubble's eye to a much older supernova remnant, the Cygnus Loop—the remains of a star that exploded 10,000 to 15,000 years ago. This supernova occurred inside a cavity of a giant star-forming region similar to the Orion Nebula. Hubble images show what happens when the supernova's blast wave slams into the nebula's material. The collision piles up thin layers of gas, each with its own temperature and electrical charge, all of which are stacked on top of each other like pancakes. "If you look at it with a ground-based telescope, all you see is a blur," says Hester. "With Hubble, you can tell that the first pancake is a buttermilk pancake and the one after that is a blueberry pancake. You can look at it in detail and compare it with the models. Overall, the observations fit pretty well with what the models predicted."

Jon Morse of the University of Colorado and his collaborators have used Hubble to image the unusual supernova remnant N132D in the Large Magellanic Cloud. Says Morse, "This is how we test how heavy elements are synthesized and how interstellar space gets enriched by these heavy elements. This is the only way to get at the stellar interior; you have to wait for the star to blow up and then you can look at the debris from the deepest layers to see what was happening." Strangely, the heavy elements in N132D are clumped together in slightly different mixtures than theory predicts. Morse questions whether the standard "onion" model can explain this particular supernova. "It's not necessarily going to uproot any theories, but it suggests that there is more to the story. There are some supernova remnants that we need to explain with different models," he says.

Morse and Hester are each part of separate teams that have used Hubble to observe a star that will one day go supernova, a powerful blue supergiant called Eta Carinae—one of the most massive and luminous stars in our Galaxy. This 100-solar-mass behemoth pumps out 5 million times more energy than the Sun. It is so bright that it can be seen with the naked eye despite its 8,000 light-year distance. In 1837 Eta Carinae flared up; by the early 1840s it had become the second brightest star in the night sky, second only to Sirius (which is only 8.6 light-years from Earth). Eta Carinae remained extremely bright for two decades before fading. Eta Carinae and other stars in its class, known as Luminous Blue Variables (LBVs for short), have been likened to geysers because they seem to undergo periodic eruptions.

Hubble images—particularly the most recent image produced by Morse—resolve amazing detail in the region surrounding this hyperactive star. Two lobes of gas are zooming away from the star in opposite directions at 1.5 million miles (2.4 million kilometers) per hour. Each lobe is about one-fourth of a light-year across and contains more than a Sun's worth of mass. When astronomers calculate the motion backward, they determine that the lobes were ejected in the 1837 flare-up. In addition, a thin but tattered disk of gas and dust is shooting away from the star's equator in the area between the expanding lobes.

Whatever caused the catastrophic eruption, the feisty star managed to survive. "This was the most cataclysmic nonterminal stellar explosion we know of," says LBV expert Kris Davidson of the University of Minnesota. While the exact cause of this eruption is still unknown, one thing is for sure: Eta Carinae will eventually explode, providing a grand spectacle for Earthlings. "It may work its way up to another outburst, or maybe in a thousand years it might just go supernova. We're not quite sure what stage the star is in or how the 1837 outburst occurred," says Morse.

Astronomers have also used Hubble to image the surface of another star destined to go supernova, the red supergiant Betelgeuse, located at the left shoulder of the constellation Orion (see photo on page 13). Betelgeuse's bloated outer atmosphere would extend past the orbit of Jupiter if the star were substituted for the Sun. Hubble resolved a gigantic hot spot in ultraviolet light, a spot 20 percent the size of the star and 200 kelvins (350 degrees F) hotter than the rest of the surface. Astronomers are still trying to figure out what causes this hot spot.

While no sane person would want to get anywhere near massive stars such as Eta Carinae, Betelgeuse, and their brethren, we can thank these stars for creating the heavy elements that make planets and life possible. The spectacular deaths of these stars spew life-giving material into the Galaxy and help initiate the birth of new stars, bringing the life cycle of stars full circle.

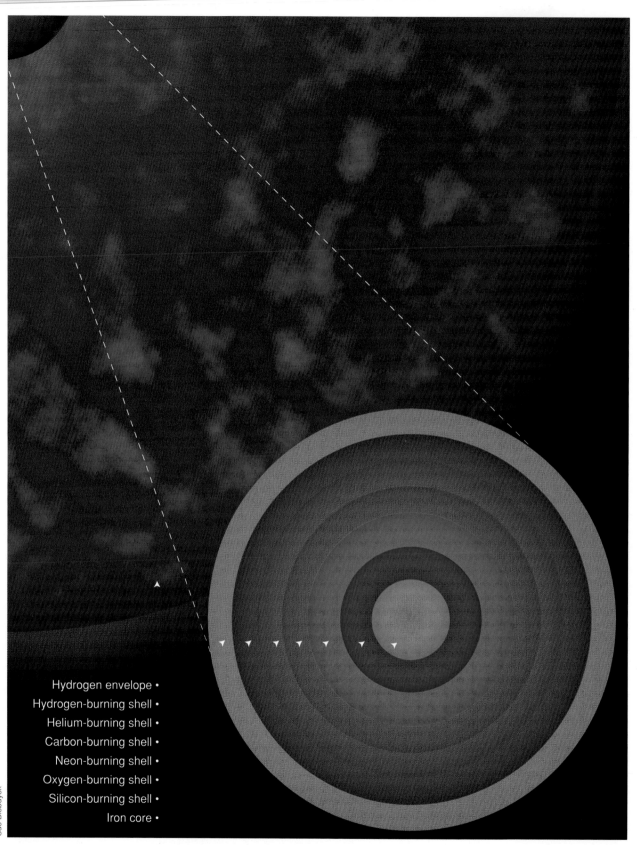

Hydrogen envelope •
Hydrogen-burning shell •
Helium-burning shell •
Carbon-burning shell •
Neon-burning shell •
Oxygen-burning shell •
Silicon-burning shell •
Iron core •

Sue Biebuyck

ONION. Right before a star explodes, its inner region resembles that of an onion. An iron core is surrounded by shells of increasingly lighter elements, all of which undergo nuclear fusion. Iron cannot be fused into heavier elements, so the core collapses under gravity and initiates a catastrophic shock wave.

Anglo-Australian Observatory/Photography by David Malin

SUPERNOVA 1987A. On February 23, 1987, the blue supergiant star Sanduleak −69°202 exploded in the Large Magellanic Cloud, the Milky Way's largest satellite galaxy. Actually, the star exploded 167,000 years earlier, but it took that long for the flash of light to reach Earth. SN1987A was the brightest supernova since the year 1604. Other stars have presumably exploded closer to home in the intervening years, but they have been obscured by dust. Stars that begin their lives with about 10 or more solar masses are destined to go supernova.

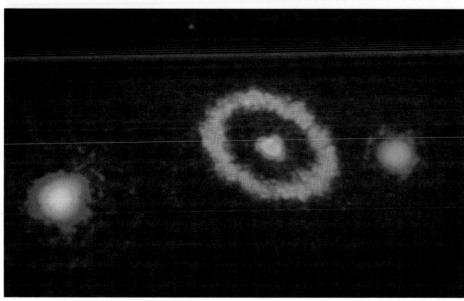

NASA

INNER RING. In 1990 Hubble became the first telescope to clearly image a ring of hot gas surrounding SN1987A. The ring is 1.4 light-years in diameter and represents material cast off by the progenitor star 10,000 years before it exploded. The pink blob is the expanding debris from the explosion, and the blue smudges are stars unrelated to the explosion. The supernova's light flash hit the ring 240 days after the explosion, causing it to glow.

Chun-Shing Jason Pun (Goddard Space Flight Center), Robert Kirshner (Harvard-Smithsonian Center for Astrophysics), and NASA

HULA HOOPS. Later Hubble images of SN1987A, taken after the Space Telescope was repaired in December 1993, revealed the same inner ring and two fainter outer rings. The outer rings are nearly three light-years in diameter and presumably represent material shed by the progenitor star 20,000 years before it exploded. The three rings lie in different planes and only appear to be stacked on top of one another because the whole system is tilted with respect to our line of sight. The central fireball has cooled to a few hundred degrees Fahrenheit. It is heated by the decay of radioactive isotopes created during the explosion.

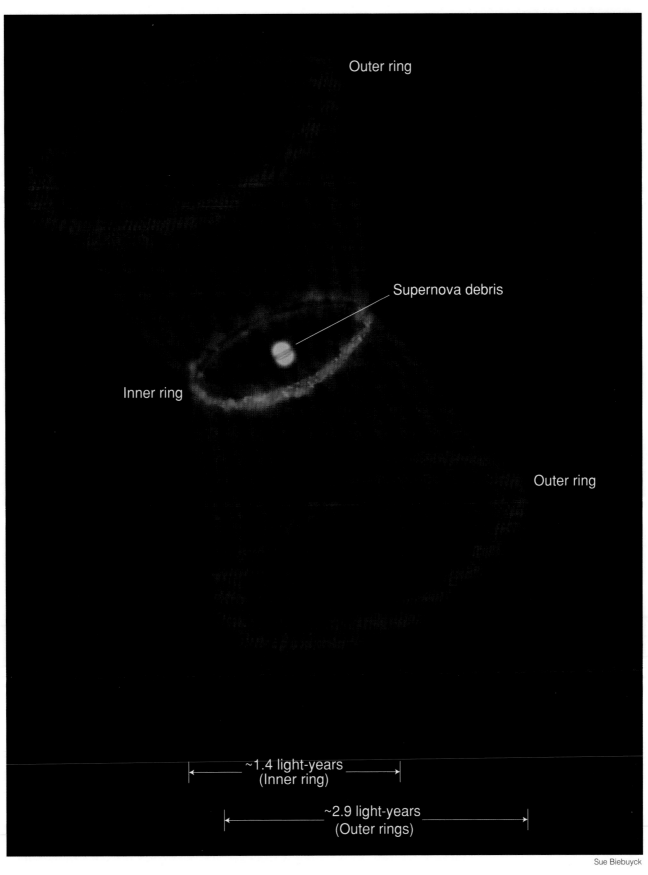

Outer ring

Supernova debris

Inner ring

Outer ring

~1.4 light-years
(Inner ring)

~2.9 light-years
(Outer rings)

Sue Biebuyck

SUPERNOVA 1987A'S RINGS probably lie on the surface of an hourglass-shaped
bubble of gas that is too faint to be seen even by Hubble.

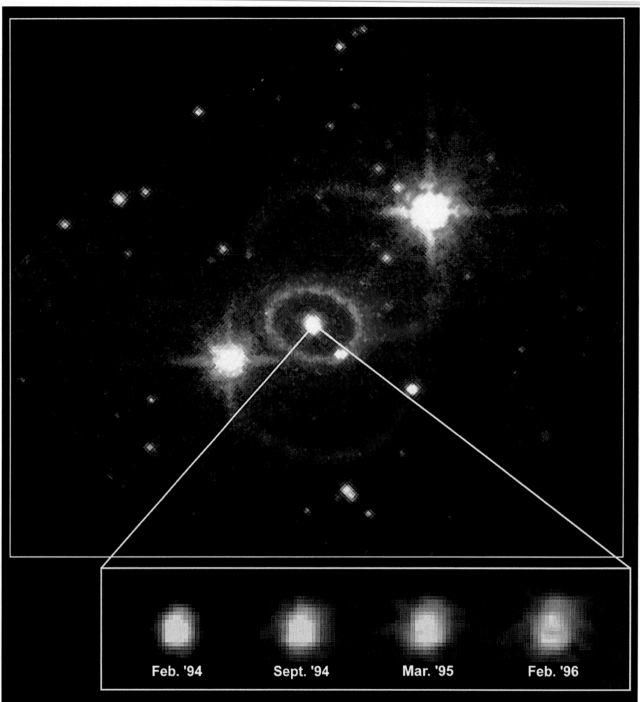

Chun-Shing Jason Pun (Goddard Space Flight Center), Robert Kirshner (Harvard-Smithsonian Center for Astrophysics), and NASA

Feb. '94   Sept. '94   Mar. '95   Feb. '96

LIGHT SHOW. This series of Hubble images (bottom) reveals the changing structure of the expanding debris from the SN1987A blast. The debris is ballooning outward at a speed of nearly 6 million miles (9.6 million kilometers) per hour (it could cross the Earth-Moon distance in less than 2$^{1/2}$ minutes) and is now about one-sixth of a light-year across. The latest image shows for the first time that the debris actually consists of two blobs moving in opposite directions. Astronomers will see the blobs smashing into the inner ring around the year 2007, causing the ring to brighten a thousandfold.

THE CRAB NEBULA is the tattered remnant of an exploded star. In 1054 A.D., Chinese astronomers recorded the supernova as a "guest star" in the night sky. The nebula is 7,000 light-years away and 10 light-years across; it shines with the intensity of 100,000 Suns. At the center of the nebula is one of nature's strangest creations: a neutron star. The neutron star crams about two Suns' worth of mass into a sphere no larger than a city. The neutron star spins 30 times per second. Astronomers on Earth see the neutron star turn on and off as its magnetic poles rotate in and out of view, like a lighthouse beam. This image was taken from the ground.

Jeff Hester, Paul Scowen (Arizona State University), and NASA

THE CRAB'S HEART. Hubble zooms into the very heart of the Crab Nebula, an eerie realm where charged particles, caught in the powerful magnetic field of a neutron star, whirl around at speeds approaching that of light. The neutron star is the lower right of the two bright stars in the center. The rapidly spinning neutron star acts like a slingshot, accelerating particles to near light-speed.

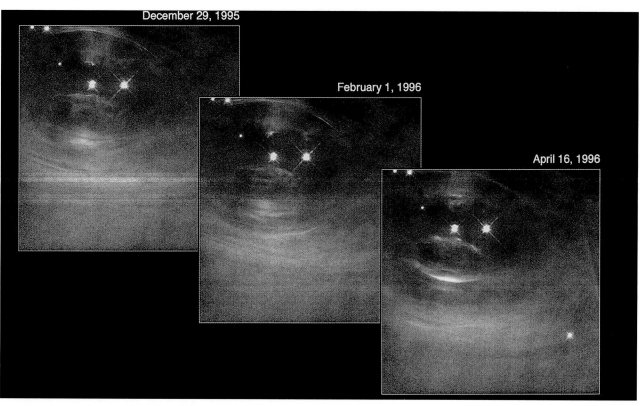

December 29, 1995

February 1, 1996

April 16, 1996

Jeff Hester, Paul Scowen (Arizona State University), and NASA

DANCING SPRITE. Three Hubble images show changes in the inner Crab Nebula over 3¹/₂ months. The neutron star is the left of the two bright stars in the upper center. Surrounding the neutron star is its equatorial wind of charged particles, which appears as light and dark concentric circles. Just above the neutron star is the "dancing sprite," a feature that comes and goes and moves around. Astronomers think a jet of particles emanating from the neutron star creates the sprite when it slams into the surrounding nebula. "There were lots of theoretical reasons to imagine that no such thing could exist, yet it does," says astronomer Jeff Hester.

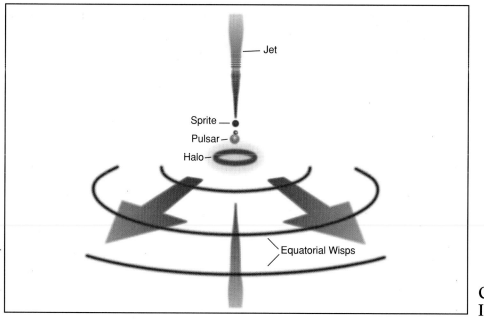

Jet

Sprite

Pulsar

Halo

Equatorial Wisps

STScI/Sue Biebuyck

CRAB NEBULA INNER REGION

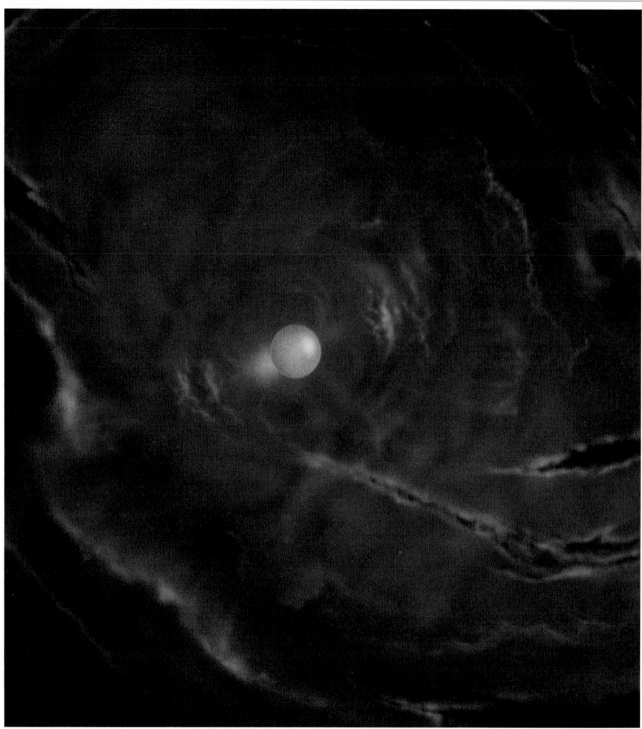

NOT A GOOD VACATION SPOT. At the very heart of the Crab Nebula lies a strange beast called a neutron star, the former core of the exploded star. This star packs about two Suns' worth of mass into a ball the size of a city. The tiny star rotates 30 times each second. Hot spots form at the magnetic poles. As the neutron star rotates, the hot spots give off pulses of radiation the way a lighthouse sends out its beams. Because this neutron star's pulses sweep past Earth, astronomers call it a "pulsar." The Crab pulsar pulses on and off 30 times each second. Electrons caught in the pulsar's powerful magnetic field buzz around the surrounding region at near light-speed.

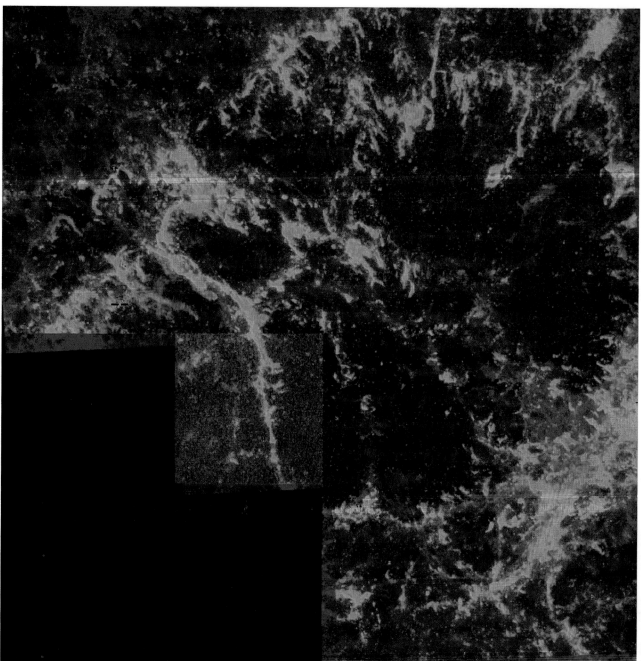

THE CRAB'S FILAMENTS. Astronomers have also used Hubble to probe the Crab Nebula's outer fringes. Electrons bound to the neutron star's magnetic field push against the star's ejecta, creating instabilities that concentrate the material into filaments. This composite image consists of two images taken through different filters and later combined. The telescope was at slightly different roll angles when these two images were taken, so the two frames are slightly offset from each other. Like most Hubble images, the colors were chosen to highlight important scientific details, not to produce natural color.

THE CYGNUS LOOP, also known as the Veil Nebula, is the tattered remains of a star that went supernova 10,000 to 15,000 years ago. This is a ground-based image of this relatively old supernova remnant, which is located about 2,500 light-years away. The Cygnus Loop is 107 light-years in diameter and appears five times larger than the full Moon. Even the brightest parts of the nebula are too faint to be seen with the naked eye, although one can see them with binoculars.

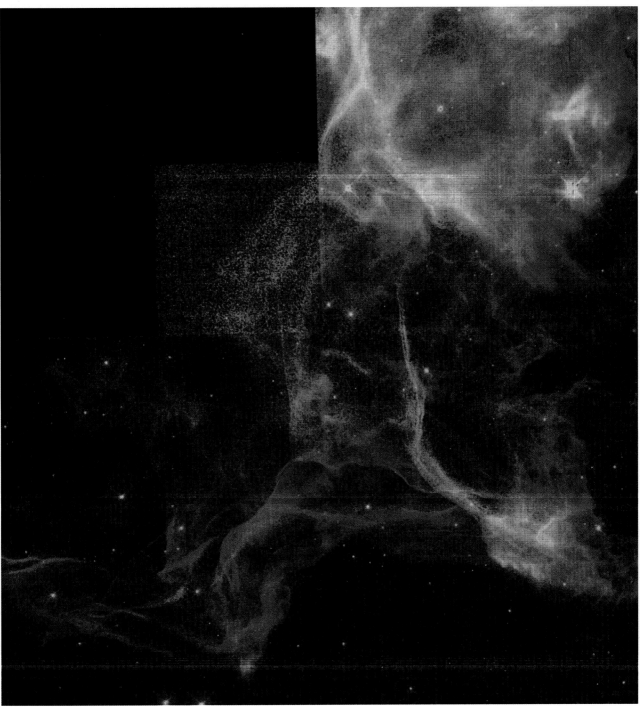

PANCAKES. This Hubble image of a small section of the Cygnus Loop shows the blast wave moving from left to right. The thin layers of gas are stacked on top of each other like pancakes. Hubble's amazing resolution allows astronomers to determine the temperature and electrical charge of each individual layer.

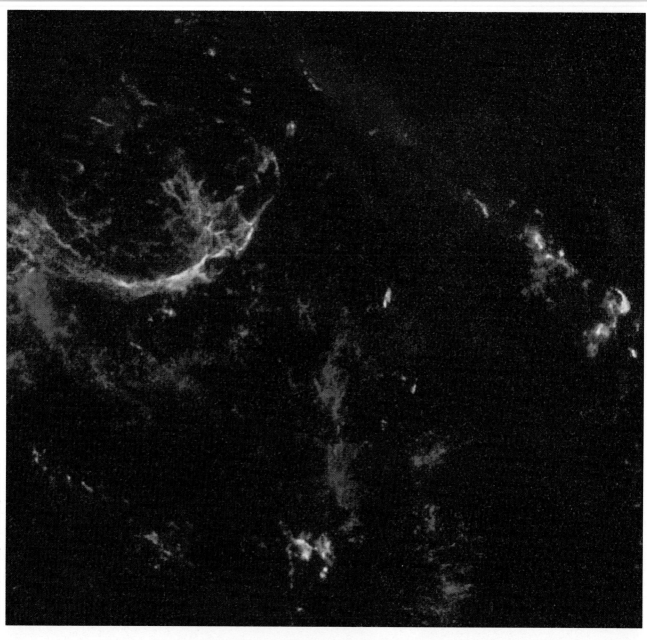

Jon Morse (University of Colorado) and NASA

STAR GUTS. Hubble images the 3,000-year-old supernova remnant N132D, located 169,000 light-years away in the Large Magellanic Cloud. The blue-green filaments represent oxygen-rich gas created by the progenitor star and then blasted into space during the explosion. Massive stars synthesize the oxygen we breathe and the iron in our blood, and then scatter these heavy elements into space when they explode. The heaviest elements of all, such as gold, lead, and uranium, are created during the explosion process.

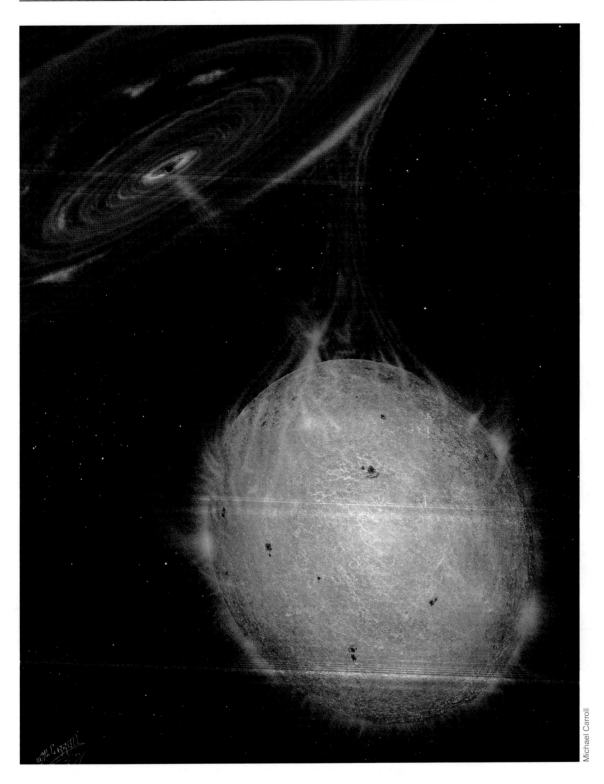

Michael Carroll

CYGNUS X-1. The most massive stars form black holes at the end of their lives. This artist's rendering depicts the black hole candidate Cygnus X-1, located some 10,000 light-years away. The black hole has a mass between 3 and 16 Suns. Its monstrous gravity draws gas from a companion 30-solar-mass blue supergiant star. The gas forms a disk around the black hole, spiraling inward toward its doom as in a whirlpool. Hubble can't see the black hole, but astronomers can detect the orbital motion of the companion star around the black hole and the X rays from the hot gas in the disk.

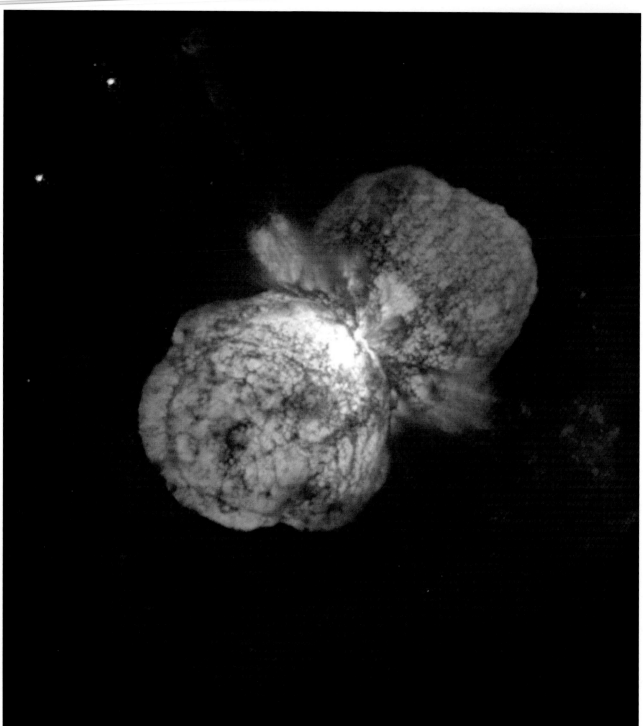

Jon Morse (University of Colorado), Kris Davidson (University of Minnesota), and NASA

ETA CARINAE may be the most massive known star, weighing in with perhaps 100 or more solar masses. The unstable blue supergiant shines with the light of 5 million Suns. This Hubble image shows twin lobes of hot gas expanding like balloons away from the star at 1.5 million miles (2.4 million kilometers) per hour. Each lobe contains about a Sun's worth of mass. The dark tendrils in the lobes are dust lanes about the width of our solar system. For some as yet unknown reason, the star ejected these two lobes in a violent outburst easily visible to the naked eye in the 1830s and '40s. Eta Carinae and other stars of its type have been likened to geysers, since they seem to undergo periodic eruptions. The star will eventually go supernova.

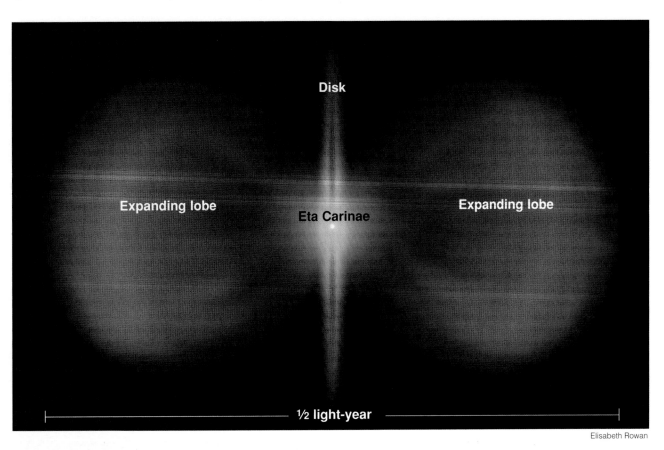

Disk

Expanding lobe

Eta Carinae

Expanding lobe

½ light-year

ETA CARINAE

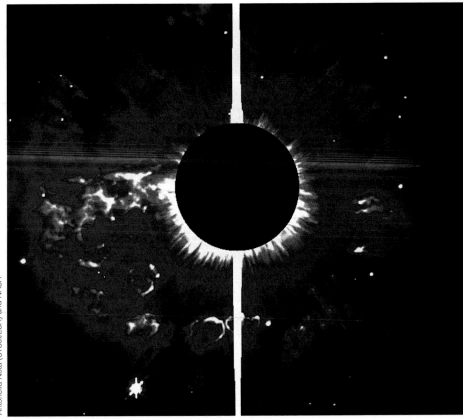

AG CARINAE belongs to the same class of stars as Eta Carinae. In this Hubble image, the star itself is blocked out so astronomers can discern details in the surrounding nebula. The bright line running from top to bottom is an artifact. The nebula, which consists of bubbles of gas and dust shed by the supermassive star, had been seen by ground-based telescopes, but it appeared as a jet rather than a group of bubbles. Astronomers use images like this one to piece together the history of the star's mass loss. AG Carinae is about 21,000 light-years away.

Andrea Dupree (Harvard-Smithsonian Center for Astrophysics), Ronald Gilliland (STScI), and NASA

BETELGEUSE is a red supergiant star about 500 light-years from Earth. Hubble took this image in ultraviolet light, the first direct image of a stellar surface other than the Sun. Betelgeuse is so humongous that if it were substituted for the Sun, its outer atmosphere would extend past the orbit of Jupiter. Hubble reveals a surface hot spot 10 percent the size of Betelgeuse itself. The origin of this hot spot, which is 200 kelvins (350 degrees F) hotter than the rest of the surface, remains a mystery. With the mass of about 20 Suns, Betelgeuse will someday go supernova.

THE BLOATED RED SUPERGIANT Betelgeuse looms large from the surface of a hypothetical planet. This is how the Sun might appear from Earth some 5 billion years from now, when it expands into a red giant. The Sun, however, will never approach the immense size of Betelgeuse.

RECYCLING MACHINES

On a clear summer night, treat yourself to one of Mother Nature's most dazzling light shows. Drive deep into the countryside, away from the polluting light of human civilization. You will see a seemingly endless band of stars stretching from horizon to horizon—the Milky Way. After pointing his crude telescope at the Milky Way in 1609, the great Italian astronomer Galileo Galilei wrote: "Upon whatever part the telescope is directed, a vast crowd of stars is immediately presented to view, many of them rather large and quite bright, while the number of smaller ones is quite beyond calculation."

Galileo didn't know it, but he was looking directly into the plane of a gigantic disk-shaped galaxy containing some 200 billion stars (that's 35 stars for each human alive today). From end to end, our Galaxy spans 100,000 light-years of space, with the Sun about 27,000 light-years from the center. Each star, including the Sun, traces its own orbit around the galactic center, which lies in the direction of the constellation Sagittarius. It takes our Sun, traveling at 150 miles (240 kilometers) per second, 230 million years to complete one circuit. The age of the dinosaurs was just getting underway the last time the Sun was on this side of the Galaxy.

The central region of the Galaxy forms a sphere that sticks a little bit above and below the plane of the Galaxy's disk, so it is often called the bulge. It contains mostly old and reddish stars. Astronomers have amassed considerable evidence that a supermassive black hole weighing in with the mass of several million Suns reigns supreme in the core.

Like many of the most beautiful galaxies in deep space, the Milky Way has a spiral form. Contrary to popular belief, the arms are not permanent features. They are places where density waves have swept through the Galaxy, piling up giant gas clouds that form new stars. The tracts of space between spiral arms contain almost as many stars as the arms themselves. But the spiral arms contain the most massive stars, those that live such short lives that they never have time to wander far from their birthplaces. These bright stars light up the spiral arms so they can be seen across the vast depths of intergalactic space. Evidence is mounting that the Milky Way is not a classic spiral galaxy, but is instead a barred spiral with a central bar about 10,000 light-years in length.

Dense spherical assemblages of stars called globular clusters also orbit the galactic center.

Globulars are almost like mini-galaxies in their own right, packing 100,000 to 1 million stars into a volume less than 100 to 150 light-years across. Most of the Milky Way's 200 or so globulars dip far above and below the disk during their orbital journeys. In general, these clusters contain ancient stars, stars that formed when the Galaxy was assembling itself 10 to 15 billion years ago. Hubble has observed a number of the Milky Way's globulars in an effort to determine exactly how old the oldest stars are. Astronomers are also peering deep into the cores of these clusters to find evidence of massive black holes.

A sea of unseen material surrounds the Galaxy in a vast spherical halo. Astronomers don't know what this so-called dark matter is, but they know that it's out there because they can detect its gravitational influence on the rest of the Galaxy. Astronomers used Hubble to see if the dark matter consisted of extremely dim stars, but the search came up empty. The exact nature of dark matter remains one of astronomy's most perplexing mysteries.

The Milky Way and its near-twin, the great spiral galaxy in the constellation Andromeda, dominate a cluster of at least 35 galaxies called the Local Group. The Andromeda Galaxy is a slightly larger version of our Galaxy. Both the Milky Way and Andromeda have their own families of small, irregular-shaped satellite galaxies. The Local Group is itself part of a much larger congregation of galaxies called the Local Supercluster.

Astronomers use Hubble to peer across the immense intergalactic void and observe star formation in other galaxies. Surprisingly, many small galaxies contain giant stellar nurseries that make the largest ones in our Galaxy seem insignificant. The Milky Way is relatively quiescent when it comes to star formation, converting perhaps five solar masses worth of gas into stars each year. But other galaxies—so-called "starburst" galaxies—furiously convert 100 or more solar masses annually. Astronomers have used Hubble's sharp vision to examine regions of intense star formation in these galaxies in an effort to determine the cause of their hyperactivity.

Galaxies such as the Milky Way have often been likened to giant recycling machines. Because the Big Bang, the stupendous explosion that gave birth to the universe 10 to 20 billion years ago, created only the three lightest elements—hydrogen, helium, and trace amounts of lithium—the first generation of stars had only these elements available. But over time, the nuclear furnaces inside stars have transformed these light elements into heavier and heavier elements. Stars spew these heavier elements into the Galaxy via their stellar winds and the occasional supernova. The amount of heavy elements has been slowly building up, and because heavy elements such as carbon, oxygen, and nitrogen seem to be essential for life, the Milky Way is gradually becoming more and more hospitable to life over time. Many of the atoms inside our bodies were forged in the interior of stars that died eons ago.

Hubble astronomers are well aware of this profound connection between the stars and ourselves. "Astronomy starts by looking up, but increasingly you turn around and look back in again," says Jeff Hester. "It's a thrill to be able to look at the Crab Nebula with Hubble, where we're actually seeing the material blasted out of that star carrying with it the alchemy that takes place inside the star—the alchemy that makes life itself possible in the universe. It starts out as a look outward for purely, 'Gee, I wonder what's out there' reasons. But in the end, it winds up providing humanity with not only a much better understanding of where we came from, but also a much better understanding of our relationship to this larger universe of which we are a part. The domain of the human mind has grown quite a lot in the last few years."

Jon Morse adds, "The sense of awe and wonder and appreciation, it's all mixed together. There are moments during your analysis where you forget about the beauty as you generate numbers. But when you first see the images, it's almost always a feeling of awe. You're saying, 'Wow, this is cool!' You are overwhelmed with all the detail you are seeing."

Paul Scowen emphasizes the need for astronomers to communicate their findings to the general public. "Astronomers need to get out there and tell the public about what it is they're learning from Hubble because it was paid for by the public. The public should get the bang for the buck that they invested in this, so we have a debt to pay there. Many of us who are working with Hubble are trying to do that sort of thing. I feel very privileged to be a part of this, because there are any number of people who could be working on this, and I just happened to be in the right place at the right time. You just want to make sure you do the best damn job that you can."

Hester adds, "The neat thing about astronomy is that there is an aspect of it that remains pure. You look at the image of the Eagle Nebula or the Lagoon Nebula, you're thinking of how it's connected with our own place in the universe. That just totally transcends any of the kinds of more mundane stories about how it comes to be. The awareness that we are gaining of the universe, the vistas that we are seeing, and the understanding that we're coming to, is just a really neat thing to be a part of."

As astronomer Carl Sagan used to say, we are literally starstuff. It's a remarkable fact that on at least one small planet orbiting an average star in an average galaxy, starstuff has evolved to a point where it can ponder and understand itself.

STRUCTURE OF THE MILKY WAY

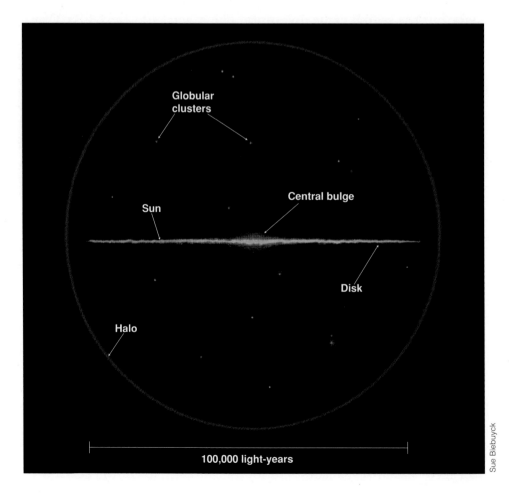

Globular clusters

Central bulge

Sun

Disk

Halo

100,000 light-years

Sue Biebuyck

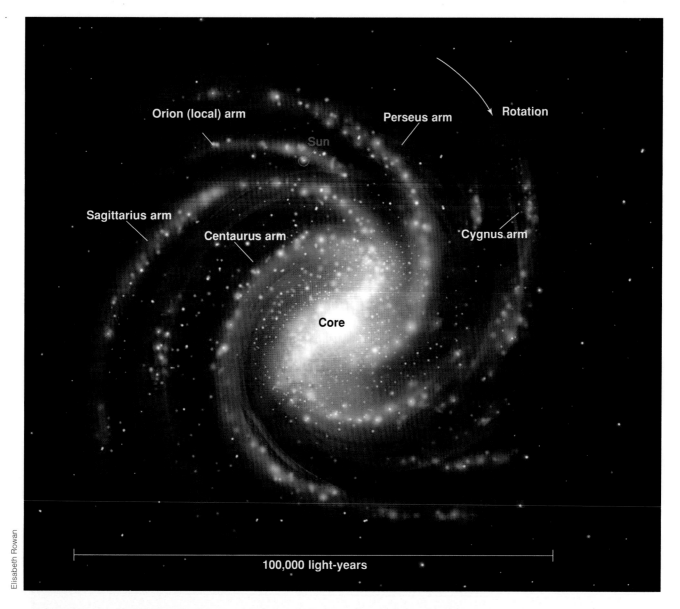

THE REAL MILKY WAY. The most recent evidence suggests that our Galaxy is a barred spiral, with a central bar about 10,000 light-years long. The graceful spiral arms are lit up by massive newborn stars. About 200 globular clusters—the densely packed spheres of stars—orbit the galactic center.

John Bahcall (Institute for Advanced Study), Don Schneider (Penn State University), and NASA

GLOBULAR CLUSTER M92 is one of about 200 globulars orbiting the Milky Way Galaxy. These stellar congregations pack 100,000 to 1 million stars into a region of space only 100 to 150 light-years across. Astronomers are using Hubble to examine the distribution of stars near the center of the cluster to see if there is evidence for a massive black hole in the core.

Puragra Guhathakurta (University of California at Santa Cruz), et al., and NASA

BLACK HOLE IN M15? Astronomers used Hubble to zoom into the heart of globular cluster M15, located 32,000 light-years away. They are tracking the motion of the innermost stars for the telltale gravitational signature of a massive black hole. The inset shows a region a mere 1.6 light-years across. This distance is less than one-third of the distance between the Sun and its nearest stellar neighbor, which illustrates how densely packed these stars are.

Deidre Ann Hunter (Lowell Observatory), et al., and NASA

NGC 1818 is a rich star cluster belonging to the Large Magellanic Cloud. The wide-field image shows a region 140 light-years across. Astronomers are studying this young cluster to get a handle on how many stars fall into various categories based on mass. The red stars are red giants and supergiants.

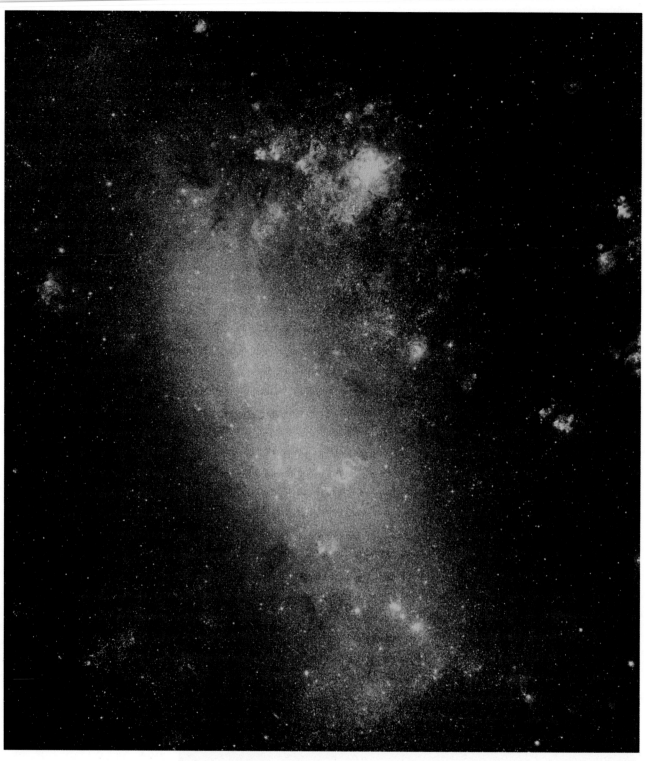

THE LARGE MAGELLANIC CLOUD is the largest of the Milky Way's dozen or so satellite galaxies. It contains roughly one-tenth the mass of the Milky Way and is located 160,000 light-years away—practically next door on the cosmic distance scale. The galaxy and its smaller sibling, the Small Magellanic Cloud, were named for the Portuguese explorer Ferdinand Magellan, who sighted the galaxies in 1520 during his crew's round-the-world voyage. The Magellanic Clouds are easily visible to the naked eye in the Southern Hemisphere. Both Magellanic Clouds will eventually be gravitationally torn apart and consumed by the much larger Milky Way.

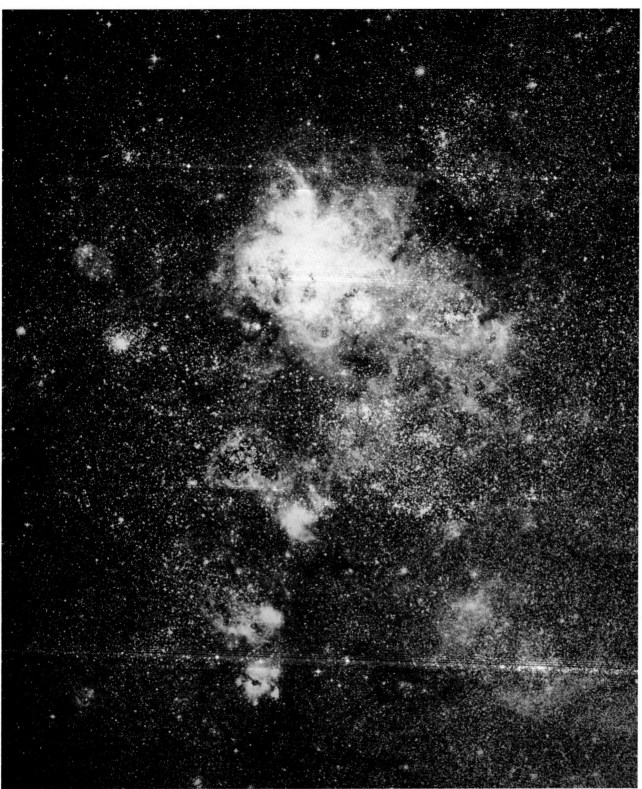

Anglo-Australian Observatory/Photography by David Malin

THE TARANTULA NEBULA, also known as 30 Doradus, is a gigantic star-forming region in the Large Magellanic Cloud. It is considerably larger than any known star-forming region in the Milky Way. If placed at the distance of the Orion Nebula, it would fill half the sky and turn night into day. The Tarantula Nebula is at the top of the photo on the facing page. (This photo and that on the facing page were taken from the ground.)

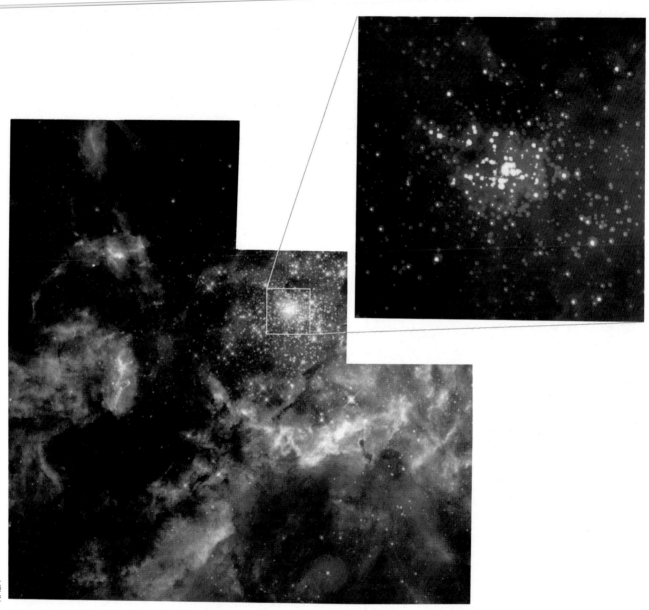

NASA

THE TARANTULA'S HEART. Hubble zoomed into the heart of the Tarantula Nebula, star cluster R136. Astronomers once thought there was only a handful of supermassive stars here, but Hubble reveals a rich cluster of more than 3,000 young stars. Most of these stars are much more massive and hotter than the Sun.

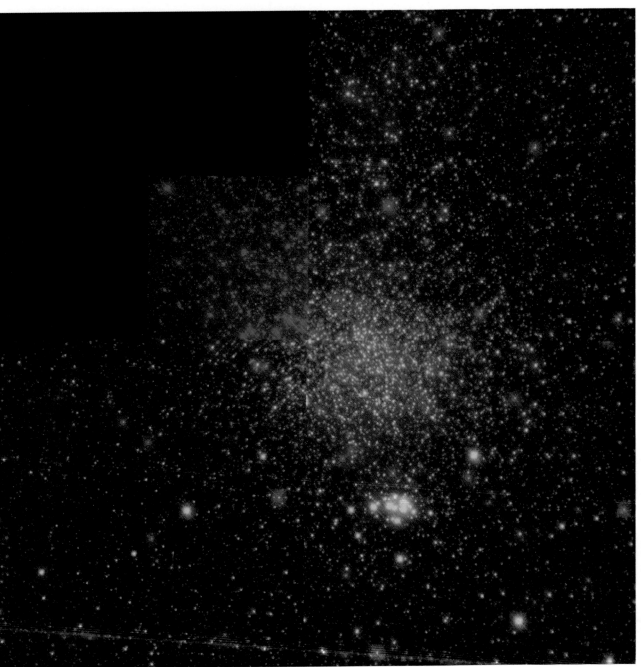

Roberto Gilmozzi (STScI/ESA/ESO), Shawn Ewald (JPL), and NASA

TWO STAR CLUSTERS appear in this Hubble image of a 130-light-year-wide region of the Large Magellanic Cloud. Most of the 10,000 stars in the field belong to the dominant cluster of yellowish stars, NGC 1850, which formed 50 million years ago inside a stellar nursery similar to the one that formed R136. The smattering of bright white stars belongs to a 4-million-year-old cluster that is more widely dispersed and some 200 light-years farther away. It's possible that supernovae in the older cluster triggered the formation of the stars in the younger cluster.

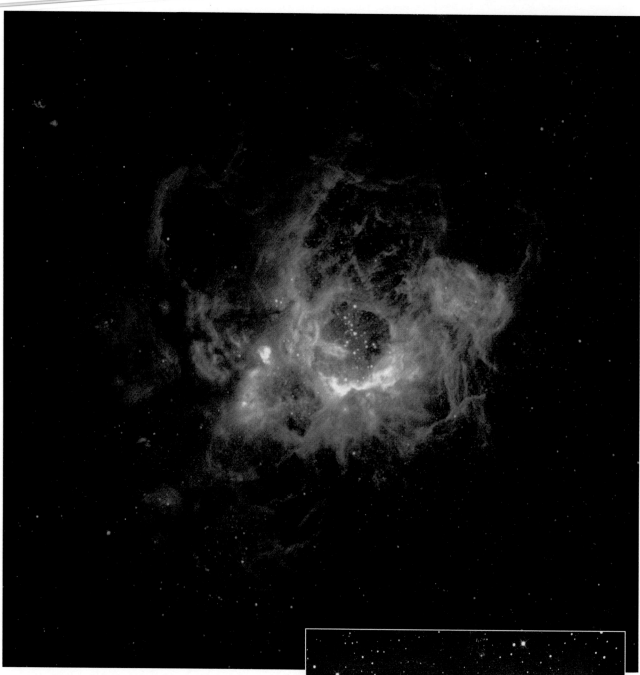

NGC 604, imaged by Hubble above, is a giant star-forming region in the small spiral galaxy M33, roughly 2.7 million light-years away. The nebula is 1,500 light-years across. Hubble was the first telescope to resolve the central star cluster, which consists of 200 massive young stars that are illuminating the nebula's interior like lanterns in a cave. Astronomers are trying to determine why a small galaxy such as M33 (seen to the right in a ground-based image, with NGC 604 boxed) can undergo such a rapid burst of star formation.

Hui Yang (University of Illinois), et al., and NASA

Tony and Daphne Hallas

Carl Weber III

THE ANDROMEDA GALAXY is the closest large galaxy, with a distance of 2.4 million light-years. Andromeda is slightly more massive than the Milky Way, and it has its own family of globular clusters and satellite galaxies. The Milky Way and Andromeda are approaching each other at a speed of a few hundred miles per second. The galaxies will collide in about 5 billion years, an event that is not as violent as one might think. Their stars are so far apart that the two galaxies will pass through each other like ships in the night without a single stellar collision. Their gas clouds will collide, however, igniting a round of furious star formation in both galaxies.

Michael Rich (Columbia University), et al., and NASA

THE ANDROMEDA GLOBULAR G1 appears as a mere blip of light in ground-based telescopes. But Hubble reveals thousands of individual stars. Astronomers can use images like this one to compare Andromeda's globular clusters to their Milky Way counterparts.

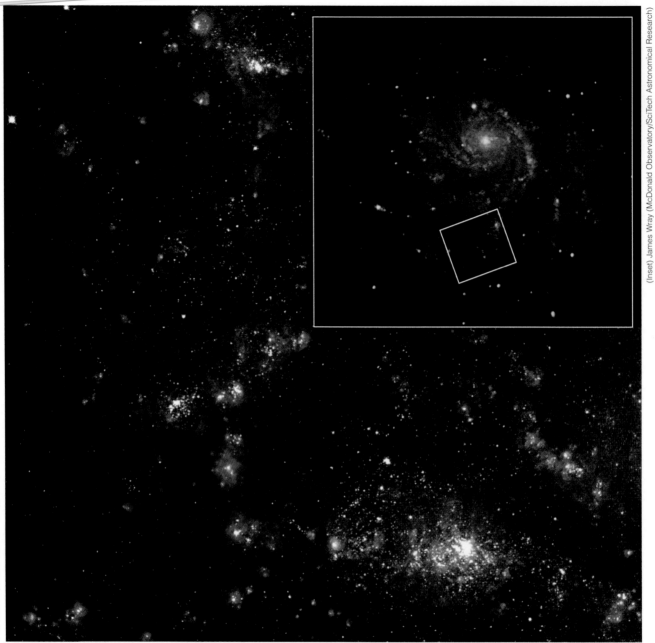

Paul Scowen (Arizona State University), et al., and NASA

(Inset) James Wray (McDonald Observatory/SciTech Astronomical Research)

STAR FORMATION IN M101. The photograph at the upper right is a ground-based telescope photograph of the spiral galaxy M101, which is located about 24 million light-years away. M101 is slightly larger than the Milky Way Galaxy, but is similar in shape. Unlike the Milky Way, M101 is undergoing extreme star formation in several regions. To learn more about the mechanisms that trigger extreme star formation in a typical spiral galaxy, and how star formation propagates throughout a galactic disk, astronomers pointed Hubble at the boxed area in the ground-based photo. Hubble reveals (the large image) giant star-forming regions in M101, which appear as bright yellowish-green patches. The largest of these star-forming regions are about the same size as the Tarantula Nebula and are being illuminated by massive young stars. Astronomers think that spiral density waves are rolling through M101 like ripples in a pond, collecting gas into giant clouds. But the star formation itself seems to be triggered by more or less random events, such as shock waves from supernova explosions.

NASA

HOME PLANET. Earth, and the amazing variety of living beings that inhabit it, are made of heavy elements forged in the nuclear furnaces of stars that died eons ago. We humans are starstuff that can contemplate the stars.

GLOSSARY

I tried to keep this book as simple as possible, but here are a few terms that some of you might not be familiar with.

Atom: the fundamental unit of matter; consists of protons, neutrons, and electrons.

Atomic nucleus: the central region of an atom; consists of protons and neutrons.

Billion: 1,000,000,000 (nine zeroes).

Binary star: a system of two stars that orbit a common center of gravity.

Black hole: a region of space where the pull of gravity is so powerful that not even light can escape; black holes can form either from the collapse of extremely high-mass stars or in the cores of galaxies.

Brown dwarf: an object that forms like a star but with a mass too small to sustain nuclear fusion in its core.

Core: the central region of a planet, brown dwarf, star, or galaxy.

Density wave: a wave of higher than normal density that propagates through a galaxy, causing stars, gas, and dust to bunch up in spiral arms, triggering the birth of new stars.

Electron: a subatomic particle with a negative electric charge; electrons surround the atomic nucleus and are much less massive than protons or neutrons.

Element: a fundamental unit of matter; consists of an equal number of protons and electrons, although the number of neutrons can vary.

Galaxy: a gravitationally bound assemblage of millions or billions of stars.

Gamma rays: the most energetic form of light.

Globular cluster: a roughly spherical congregation containing hundreds of thousands of stars.

Gravity: the attractive force that all objects exert on one another; the greater an object's mass, the stronger its gravitational pull.

Helium: the second lightest element; consists of two protons, and usually two neutrons and two electrons; about 8 percent of the atoms in the universe are helium.

Hydrogen: the simplest and lightest element; consists of just a single proton and electron; about 90 percent of the atoms in the universe are hydrogen.

Infrared radiation (also known as infrared light): a form of light with slightly lower energy than humans can see with their eyes.

Kelvin: degrees above absolute zero on the Celsius temperature scale; one degree kelvin equals 1.8 degrees Fahrenheit.

Light-year: the distance light travels in one year, equivalent to 6 trillion miles (10 trillion kilometers).

Luminosity: the amount of light that an object gives off.

Mass: the amount of matter contained in an object.

Million: 1,000,000 (six zeroes).

Nebula: a cloud of gas; some nebulae represent stellar nurseries, others represent stellar graveyards.

Neutrino: a subatomic particle with either no mass or nearly no mass that is produced by nuclear fusion and supernovae.

Neutron: a subatomic particle with no electric charge; it has about the same mass as a proton.

Neutron star: the collapsed, city-sized remnant of a high-mass star.

Nova: a violent explosion on the surface of a white dwarf.

Nuclear fusion: the process by which two or more atomic nuclei combine to form a heavier atomic nucleus, releasing energy in the process.

Open cluster: a system containing hundreds or thousands of stars that formed from the same stellar nursery.

Photons: individual "particles" of light.

Planetary nebula: the glowing shell of gas given off by a dying star.

Proton: a subatomic particle with a positive electric charge.

Protostar: a cloud of hot, dense gas that is gravitationally collapsing into a star.

Red dwarf: a low-mass star much smaller, cooler, and less luminous than the Sun.

Red giant: a cool star that has expanded to a size much greater than the Sun.

Shock wave: a powerful wave caused by a violent change in density, pressure, or temperature; shock waves bunch together the material they travel through.

Star: a self-luminous sphere of hot gas held together by gravity; ordinary stars generate energy by nuclear fusion in their cores.

Stellar wind: a stream of electrically charged subatomic particles given off by stars.

Supergiant: the largest and most luminous class of stars in the universe.

Supernova: the cataclysmic explosion of a star.

Trillion: 1,000,000,000,000 (twelve zeroes).

Ultraviolet radiation (also known as ultraviolet light): a form of light with slightly higher energy than humans can see with their eyes.

White dwarf: the collapsed, Earth-sized remnant of an intermediate-mass star like the Sun.

INDEX